Circles

A Mathematical View

Originally published in 1957 by Pergamon Press.
A corrected and enlarged edition was first
published in 1979 by Dover Publications, Inc.

© 1995 by
The Mathematical Association of America (Incorporated)
Library of Congress Catalog Card Number 95-77405

ISBN 0-88385-518-6

Printed in the United States of America

Current Printing (last digit):
10 9 8 7 6 5 4 3 2 1

Circles

A Mathematical View

Dan Pedoe

Published by
THE MATHEMATICAL ASSOCIATION OF AMERICA

SPECTRUM SERIES

The Spectrum Series of the Mathematical Association of America was so named to reflect its purpose: to publish a broad range of books including biographies, accessible expositions of old or new mathematical ideas, reprints and revisions of excellent out-of-print books, popular works, and other monographs of high interest that will appeal to a broad range of readers, including students and teachers of mathematics, mathematical amateurs, and researchers.

Mathematical Association of America
1529 Eighteenth Street, NW
Washington, DC 20036
800-331-1MAA   FAX 202-265-2384

PREFACE

It gives me great pleasure that the Mathematical Association of America is republishing my book on circles, and I have been happy to adopt their suggestion that I write an introductory Chapter Zero to introduce and discuss the geometrical terms which may not be familiar to a new generation of students. This additional chapter is based on lectures I have given in India and Australia entitled "Geometry and the Pleasure Principle," and I hope that my enthusiasm may be shared by others. I cannot think of any possible moral disapproval which can be held to be valid if one enjoys geometrical activity, and perhaps one of the beliefs of my youth that excellence in geometry leads to excellence in other branches of mathematics may once again become current.

For this criterion to be tested, of course, geometry must not be excluded, as it has been for too many years, from school and college curricula. Fortunately there are many signs of a revival of interest in geometrical teaching, and the MAA is one of the leaders in this very heartening activity.

Boldface numerals in parentheses, such as (**n**) which appear in Chapter Zero refer to the list of references which appear at the end of Chapter Zero. These articles and books can be consulted, of course, for further and more detailed information.

D. P.
Minneapolis, Minnesota
June, 1994

CONTENTS

CHAPTER III

CHAPTER IV

CHAPTER 0

Geometry is both an art and a science, and its pursuit can give great pleasure. The only tools needed are paper, a pencil, a simple school compass (sometimes called a *pair* of compasses), a ruler to draw straight lines connecting pairs of points, and perhaps a protractor to measure angles in degrees. The protractor reveals at once the great antiquity of the subject, for the markings around the semicircle are in multiples and divisions of 60°, and this system of numeration dates back to Babylonian times. We note that the angle on a straight line is 180°, and that a right angle is 90°.

We are told that the Greeks of the third century BC, whose geometrical learning was collated by Euclid, were not able to fix the arms of the compass they used at a constant inclination, so that as the instrument was raised from the sand tray on which they made their drawings, the arms collapsed together. Our modern compass has arms which can be made to stay at a fixed distance apart, and we can immediately draw that most æsthetic of all curves, a circle. Marking a point on this circle, we can "step around" the circle (Figure 1^0) and find the point which is at the other end of the diameter, a property we shall use later (p. xix) and, drawing more circles, we can produce a much-loved pattern, the basis of the windows found in many cathedrals (Figure 2^0).

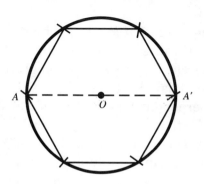

FIG. 1^0

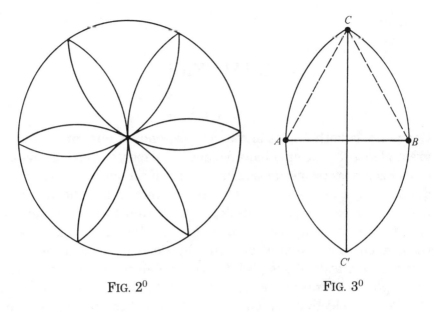

FIG. 2^0 FIG. 3^0

Our compass with the fixed radius AB (Figure 3^0) can produce the vertex C of an equilateral triangle ABC, in which $AB = BC = CA$, and if we assume, as is the case in Euclidean geometry, that the angles of any triangle add to 180°, it is clear that the angles of an equilateral triangle are each equal to 60°.

If we allow the circular arcs CAC', CBC' to remain in our figure, we have constructed a shape regarded as fundamental in mediæval architecture, the *fish bladder*. But of even greater importance is the fact that we have a method, using our compass and ruler, of constructing the perpendicular bisector CC' of a given segment AB.

Symmetry suggests that if P is any point on CC', the distance PA is equal to the distance PB. We use this fact to construct the *circumcircle* (the circle which passes through the vertices A, B, and C) of any given triangle ABC.

We draw the perpendicular bisectors of AB and BC. If these lines intersect at O, then $OA = OB$ (Figure 4^0), and also $OB = OC$, and we deduce, subconsciously using one of Euclid's postulate-axioms, that $OA = OC$, and finally, although we have not proved this converse, that O must lie on the perpendicular bisector of CA.

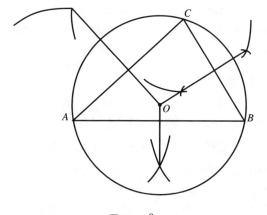

FIG. 4^0

A circle with radius $OA = OB = OC$ and centre O will pass through the vertices A, B, and C of the triangle ABC. This point O is called the *circumcentre* of the triangle ABC.

There are more circles conventionally associated with a triangle ABC, one *incircle* and three *excircles,* and to introduce these we must talk about the *angle* between two given lines.

If A is a point on one line, B on the other, and O the intersection of the lines, we can call OA and OB *half-lines,* and use our protractor to measure the angle between them. If the half-lines coincide, the angle is zero degrees ($0°$), and if OB is just an extension of OA (or rather AO), the angle between the lines is $180°$, that is: two right angles.

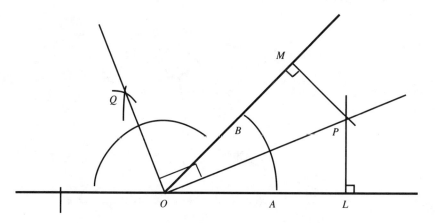

FIG. 5^0

To draw the line bisecting the angle AOB, we use our compass and mark $OA = OB$ on the respective lines, and then find P where $AP = BP$, the intersection of arcs centres A and B respectively. Joining P to O, we have a line which clearly bisects the angle AOB (Figure 5^0).

When we say *clearly,*, we are using a theorem that states that if in two given triangles OBP and OAP we have the equalities

$$OP = OP, \quad OA = OB, \quad \text{and} \quad AP = BP,$$

then the triangles are *congruent* (identical, in a sense), and the corresponding angles are equal, so that $\angle BOP = \angle AOP$.

If we draw the *external* bisector of angle AOB, the line OQ, then it is not surprising to see that OQ is at right angles to OP. The proof depends on the simplest algebra.

Now we can draw a circle which *touches* the sides of a given triangle ABC. A line *touches* a circle at a given point on the circle if it passes through the point and has no further intersection with the circle. Such a line is *tangent* to the circle at the given point, and is at right angles to the line drawn from the centre of the circle to the given point. You may think that this does not need proof. Assuming this fundamental theorem, we "drop a perpendicular" PL from P onto OA, and a perpendicular PM onto OB, and we have the equality $PL = PM$. A circle centre P and radius $PL = PM$ will touch both OA and OB (Figure 5^0).

To construct the *incircle* of a given triangle ABC, we draw the bisector of angle A, and the bisector of angle B, and let I be the intersection of these two lines (Figure 6^0).

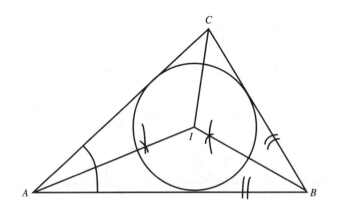

FIG. 6^0

For all points on AI the perpendicular distance of P from AB is equal to the perpendicular distance of P from AC, with a similar argument holding for points P on the bisector of angle B. Hence the point I is equidistant from AB, AC, and BC, and a circle centre I can be drawn to touch AB, AC, and BC. The point I is the *incentre* of triangle ABC, and we have the theorem: *the bisectors of the angles A, B, and C meet at a point.*

We show how to draw the excircle opposite C by drawing the bisectors of the *external* angles of the triangle at A and B, respectively. This excentre lies on the *internal* bisector of angle C. The triangle has one incircle and three excircles (Figure 7^0).

There are two other familiar points connected with a triangle ABC. One is the intersection of the *medians,* the lines joining the vertices of the triangle to the midpoints of opposite sides. This point M is often called the *centroid* of the triangle, and trisects each median (Figure 8^0).

A final point in our brief discussion is the *orthocentre H* of a triangle ABC. If we drop perpendiculars from the vertices onto

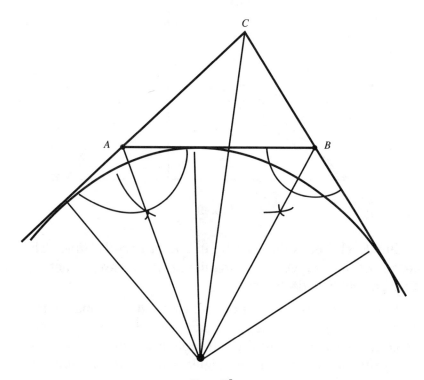

FIG. 7^0

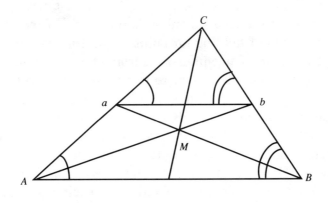

FIG. 8^0

the opposite sides of triangle ABC, these three lines will pass through the same point H. If one angle of the triangle exceeds a right angle, the point H will lie outside the triangle. Curiously enough, this point is not mentioned in Euclid (Figure 9^0).

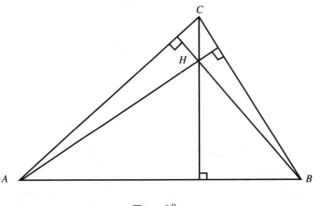

FIG. 9^0

But Euclid does discuss the notion of *parallel lines*. The famous Postulate V, which distinguishes Euclidean geometry from other geometries, declares:

> If a straight line falling on two straight lines makes the interior angles on the same side less than two right angles, the two straight lines, if produced indefinitely, meet on the side on which the angles are less than two right angles (Figure 10^0).

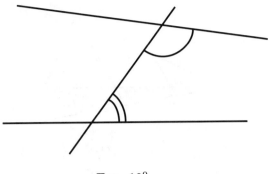

FIG. 10^0

If we now define *parallel* lines as straight lines which, being in the same plane and being produced indefinitely in both directions, do not meet one another in either direction, we obtain the equal angles shown in Figure 11^0.

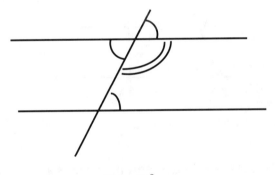

FIG. 11^0

The interior angles on the same side of the straight line falling on the two parallel lines add to two right angles.

From these preliminaries it follows rapidly that in Euclidean geometry the angle sum of any proper triangle (whose vertices are not collinear) is two right angles, and we move on to the notion of *similar* triangles. These are triangles of the same shape (Figure 12^0).

In the figure, angle a is equal to angle A, angle b to angle B, and angle c to angle C. It can be shown that in such a case the sides are proportional: that is, if $BC = kbc$, then $CA = kca$, and $AB = kab$. Conversely, if equalities of this type hold between the

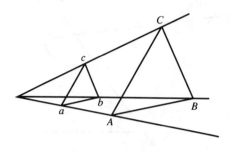

FIG. 12^0

sides of two given triangles, then these triangles have the same shape (they are *similar*), and corresponding angles are equal.

Our final essential theorem before we move on to properties of circles refers to transversals of three given parallel lines (Figure 13^0).

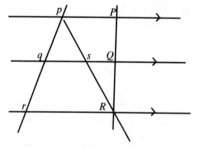

FIG. 13^0

If one transversal is pqr and another is PQR, then
$$pq/qr = PQ/QR.$$
If we join p to R, we have two triangles with a common side, and in the one:
$$pq/qr = ps/sR,$$
and in the other
$$ps/sR = PQ/QR.$$
We note that in the figure showing the concurrence of the medians of a triangle (Figure 8^0) $AB = 2ab$, and ab is parallel to AB.

Before we examine circles in some detail, let us dally with a lovely problem which captured my attention many years ago.

Given two points O and A, and only a compass, how can one determine, exactly, the midpoint of the segment OA?

This problem came from some Cambridge University examination papers (called the Mathematical Tripos) in the days when the problems had to be solved by the methods of Euclidean geometry, the calculus being strictly forbidden. The stress is on the word "exactly." One is tempted to doodle and guess . . . .

However ... draw the circle centre A and radius AO, and, beginning at O, step round this circle to B, the opposite end of the diameter through O (see Figure 1^0). With O as centre draw the circle through A, and with B as centre and radius BO draw a circle which cuts the circle centre O in P and Q. With P as centre and radius PO, draw a circle, and with Q as centre and radius PO draw another circle. These last two circles intersect in O and the desired midpoint M.

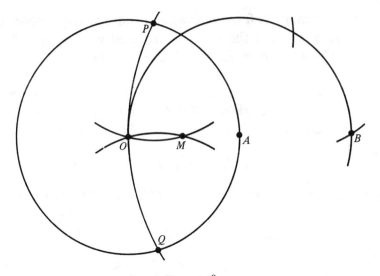

FIG. 14^0

The proof of the preceding construction depends, very simply, on the theory of similar triangles. There are two isosceles triangles (triangles with two equal sides) in the figure, and the base angles being equal, the sides are proportional, and this gives the result. This problem occurs again in the main section of this book (p. 24).

(The two isosceles similar triangles are POM and BPO.) In the main section of this book it will be shown that the ruler can be dispensed with in the constructions of Euclidean geometry, except, of course, for drawing an actual line connecting two points.

How does one check that a given ruler does draw a straight line?

Take two points A and B, and use the ruler to draw the alleged straight line AB. Now turn the ruler on its other side, and again draw the line AB. If the two lines you have drawn connecting A to B enclose a space, your ruler is defective. Since rulers traditionally receive a lot of wear, it may pass this test, but may still have notches in its edge.

It is easier to draw an accurate circle than it is to draw a straight line, and so one should not be too surprised to see that books have been written with the title: *How to draw a straight line.*

A final remark before we examine some of the properties of circles. There are two theorems which deal with a triangle and lines, the Ceva theorem, and the Menelaus theorem.

Ceva's theorem. *If lines AP, BP, and CP through the respective vertices of a triangle ABC intersect the respective opposite sides in the points L, M, and N, then (Figure 15⁰)*

$$(BL/LC)(CM/MA)(AN/NB) = 1.$$

If this relation holds, the lines AL, BM, and CN meet in a point P.

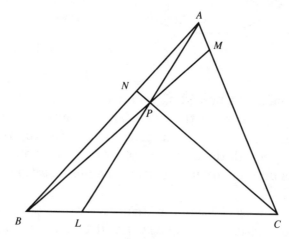

FIG. 15⁰

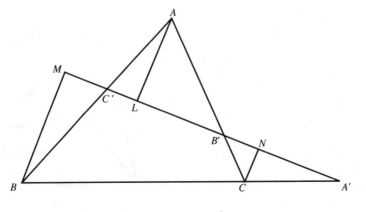

FIG. 16^0

Menelaus's theorem. *If a line intersects the respective sides BC, CA, and AB of a triangle ABC in the points A', B', and C', then (Figure 16^0)*

$$(BA'/A'C)(CB'/B'A)(AC'/C'B) = -1.$$

If this relation holds, the points A', B', and C' lie on a line.

The Menelaus theorem is proved simply by dropping perpendiculars AL, BM, and CN onto the line, and using the ratio properties of similar triangles, and the Ceva theorem is proved by comparing the areas of triangles with a vertex in common, and along the same baseline. The lengths involved in both theorems are *signed* lengths. The time-span between the appearance of the Menelaus theorem and the Ceva theorem runs to hundreds of years. (For further information, see **1**.)

The circle has some beautiful intrinsic properties. If O is the centre of a given circle, P and Q are given points on the circle, and the point R is a variable point on the circle, then the angle PRQ is a constant angle as R moves on each of the parts of the circle bounded by the points P and Q. The sum of these two constant angles is $180°$ (Figure 17^0).

This theorem is an obvious deduction from the theorem that the angle POQ is twice the angle PRQ. Since the angle on a line is 180^0, and the sum of the angles of a triangle is also 180^0, the external angle of a triangle is the sum of the two interior and opposite angles, and if, as in the case we are considering, the triangles POR and QOR are both isosceles, the sides OR

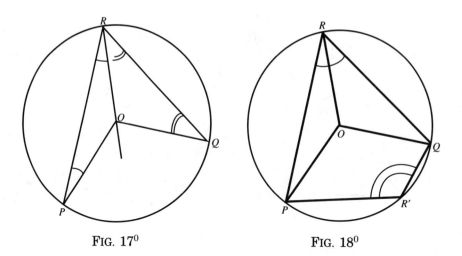

and OP being equal, and also the sides OR and OQ, the angles shown in the figure are equal, and so angle POQ is twice angle PRQ.

The above proof has to be modified when PR crosses OQ, but the theorem still holds.

However, if R lies on the other part of the circle bounded by P and Q (Figure 18^0), the angle POQ is now the complement of the angle with R on the first arc, the two angles adding to $360°$, and we have the theorem: *If $PR'QR$ are points in this order on a circle, then*

$$\text{angle } PRQ + \text{ angle } PR'Q = 2 \text{ right angles.}$$

The converse of this theorem also holds for points $PR'QR$ which form a convex quadrangle. If the two angles shown in Figure 18^0 add to two right angles, then the four points are *concyclic*, lying on a circle.

Another potent theorem which holds for a circle leads to the *power* concept. If P is a given point and lines PAB, $PA'B'$ through P intersect a given circle in the points A, B, and A', B' respectively, then

$$PA \cdot PB = PA' \cdot PB'.$$

This number, which depends only on P and the given circle is called the *power* of the point P with respect to the given circle. The proof of this theorem (Figure 19^0) is immediate from the recognition that the triangles PBA' and $PB'A$ are similar, so

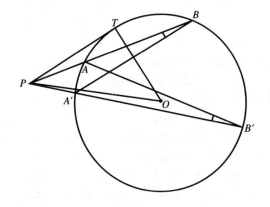

FIG. 19^0

that

$$PB/PA' = PB'/PA.$$

If P lies outside the circle, and PT is a tangent at T, the power of P is also $(PT)^2$, and if O is the centre of the circle, R its radius, this is equal to $(PO)^2 - R^2$, by the Pythagorean theorem.

When P lies inside the circle (Figure 20^0) we again have similar triangles, and a similar proof. But now we cannot draw a tangent from P to the circle. If, however, PCC' is at right angles to PO, then $PC = PC'$, and so the power is equal to $(PC)^2$, which is equal to $R^2 - (PO)^2$. It is convenient to write the power as $(PO)^2 - R^2$ in this case also, a full justification arising when we discuss circles using analytical geometry, as we do later in this

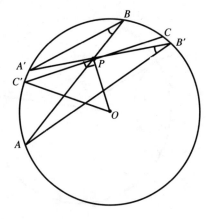

FIG. 20^0

book. Of course, the power of P is zero when P lies on the circle, and it is positive when P lies outside the circle, and negative when P lies inside the circle.

We now come to the important concept of *orthogonal* circles, circles which cut each other at right angles. Of course, circles can cut each other at other angles, the angle of intersection being measured by the angle between the tangents to each circle at a point of intersection. But orthogonal circles are especially simple to construct. Draw the tangent at a point T on a circle, and on the tangent choose a point O. With O as centre, draw the circle with radius OT (Figure 21^0). We now have two circles which have tangents at the point T of intersection at right angles. Of course, this is true for the other point of intersection also.

Now let QOP be a diameter of the circle centre O which intersects our first circle in the points A and B. Then the power of O with respect to the first circle is $OB. OA$, and this is also $(OT)^2$. But OT is a radius of the second circle, so that we have

$$OB \cdot OA = (OP)^2 = (OQ)^2. \qquad (1)$$

The set of points, consisting of the pair A, B, and the pair P, Q, is called an *harmonic* set, and has special properties, as we shall show. A and B are said to be *harmonic* conjugates with respect to P and Q, and P and Q are said to be harmonic conjugates with respect to A and B. P divides the segment AB internally in the same ratio that Q divides it externally. That is

$$AP/PB = -AQ/QB.$$

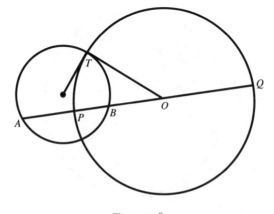

FIG. 21^0

Likewise
$$PA/AQ = -PB/BQ.$$
Each of these equalities is rapidly deducible from the other.

We now show that from $AP/PB + AQ/QB = 0$, we can deduce the equations (1) above, and conversely. We use coordinates for simplicity, taking the origin at O, the midpoint of PQ.

We have the equation
$$(p - a)/(b - p) + (q - a)/(b - q) = 0,$$
where $p + q = 0$, and substituting for q,
$$(p - a)/(b - p) + (-p - a)/(b + p) = 0.$$
Multiplying by $(b - p)(b + p)$, which is not zero, we obtain
$$(p - a)(b + p) - (p + a)(b - p) = 0,$$
which simplifies to
$$2p^2 - 2ab = 0, \quad \text{that is} \quad OA \cdot OB = (OP)^2 = (OQ)^2.$$

From this final equation we can reverse every step taken, and arrive back at
$$AP/PB + AQ/QB = 0.$$
Harmonic sets will play an important part in our next chapter.

We have two more theorems to discuss before returning to the pleasures of geometrical constructions, both involving the orthocentre H of a triangle ABC.

If AH intersects BC in P, and the circumcircle in K, then $HP = PK$ (Figure 22⁰).

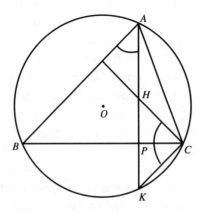

FIG. 22⁰

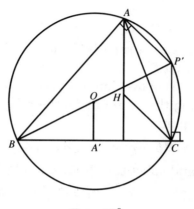

FIG. 23^0

From the figure the proof is clear. The two triangles CPH and CPK are congruent, both have right angles at P, they have the common side PC, the angle PCK is equal to the angle BAK, subtended on the circumcircle by BK, and angle BCH is equal to angle BAH, both differing from a right angle by the angle B.

If A' is the midpoint of BC, then $AH = 2OA'$ (Figure 23^0).

If the perpendicular at C to BC intersects the circumcircle again at P', then $P'B$ is a diameter of the circumcircle, and so must pass through the circumcentre O. Since $P'AB$ is also a right angle, $P'A$ is parallel to CH. Hence $P'AHC$ is a parallelogram, and AH is equal to $P'C$. Finally, in the right triangle $BP'C$, OA' bisecting both BP' and BC, we have $P'C$ is equal to $2OA'$, and so AH is equal to $2OA'$.

We now return to geometrical constructions which produce pleasing and unexpected figures. To start with, we draw a circle, which we shall call the base circle. Fix a point A on this base circle, and then draw circles which have their centres on the base circle, but always pass through the given point A. As you move systematically around the base circle, the drawn circles change in size and you will produce, as an envelope, a curve touched by all the circles you have drawn, a heart-shaped curve, a *cardioid,* (Figure 24^0).

At the special point A, two parts of the curve touch. Such a point is called a *cusp*. This curve has been known for several centuries (see (**2**)). You may want to experiment, moving the point A off the base circle.

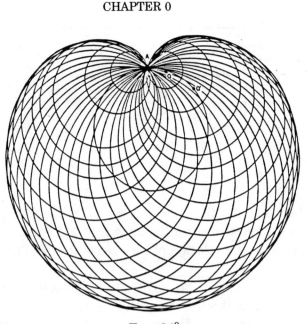

FIG. 24⁰

Curves are usually regarded as traced out by points on the curve considered, but they can also be defined by their tangents. A simple method for doing this is to *crease paper.*

We begin with a sheet of paper on which we designate one edge as ℓ, (Figure 25⁰), and mark a point on the paper, which we call S. Fold the paper so that the edge ℓ passes through S, and then crease the paper along a line so that the edge ℓ continues to pass through S. (Plain writing paper will do, tissue paper is too

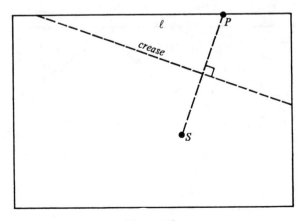

FIG. 25⁰

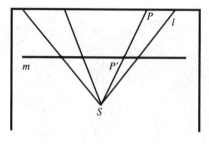

FIG. 26^0

thin, cardboard is too thick.) When we open the sheet of paper
out again, we have the situation shown in Figure 25^0, and it is
important to note the relation of the crease to the edge ℓ and the
point S.

    The crease bisects SP at right angles, P being the point on
the edge ℓ where S landed when the paper was folded. If P' is
the point where the crease meets SP, what kind of curve does P'
trace as P moves along the edge ℓ? Can you prove that P' traces
a line parallel to ℓ, and halfway between S and ℓ? See Figure 26^0.

    If you continue making creases, you will see that they touch
an attractive curve, the *parabola,* with S as its focus. I say *the*
parabola, all parabolas have the same shape, but obviously the
size of our parabola depends on the distance of S from ℓ.

    Noting the relation of the crease to SP in Figure 27^0, and
drawing the path (the *locus*) of P', we can draw tangents to the

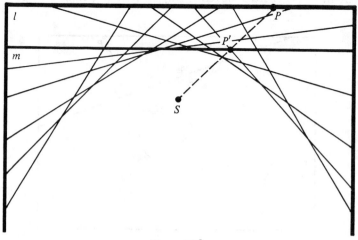

FIG. 27^0

parabola without making creases by using a simple geometrical instrument, a right triangle, made of wood (or plastic), called a *setsquare* in England, but not in America. Let one side of the right angle pass through S, the vertex of the right angle lying on m, the halfway line, and draw the other side of the right angle.

This is, of course, a quicker method than paper creasing. The halfway line m is a tangent to the parabola, we note, and the curve is symmetrical about the perpendicular from S onto m. This line is called the *axis* of the parabola, and m is the tangent at the *vertex* of the parabola.

It is not difficult to obtain the *focal* properties of the parabola from this construction via its tangents (see (**2**)), and such properties were known to Albrecht Dürer (1471–1528), a great artist and geometer (Figure 28^0). These properties are used in car headlamp reflectors, in radio antennas, and so on. Parabolic dishes are used everywhere.

What happens if we pursue our paper-folding-creasing construction, and choose a circle C, and mark a point S inside it,

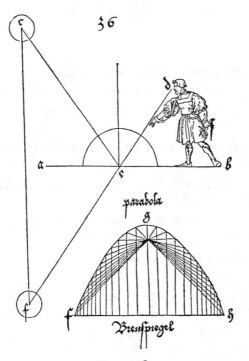

FIG. 28^0

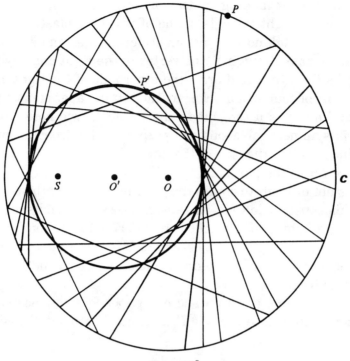

FIG. 29^0

and then fold our paper so that a point on the circle C itself falls
on the point S when we crease the paper. Our creases produce
the tangents to an ellipse (Figure 29^0), with the point S as one
focus.

We can again draw lines instead of making creases if we note
that if P is the point on the circle C which falls on S, then the
midpoint P' of SP moves on a circle with radius half that of the
circle C. See Figure 30^0.

Take O as the centre of C, and let O' be the midpoint of SO.
Prove that $O'P'$ is equal to $OP/2$, and is therefore a constant
length. This follows from the fact that P' bisects SP and O' bisects SO.

To obtain tangents to an ellipse we therefore join S to points
P' on the half-circle, and draw lines at right angles through the
points P' to give the creases. The shape of the ellipse depends
essentially on the ratio of the distances of S from O and the end
of the radius of C which passes through S.

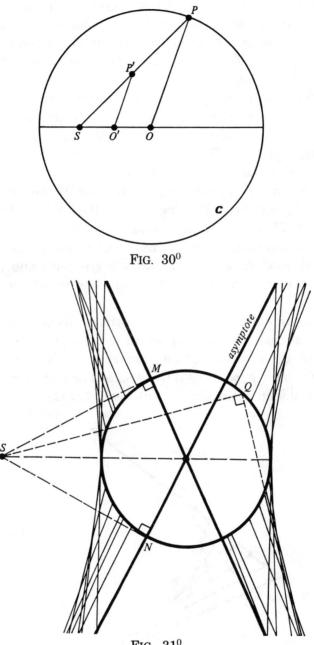

FIG. 30^0

FIG. 31^0

Finally, let S be a point outside the circle C, and repeat the folding-creasing construction. We obtain both branches of an hyperbola (Figure 31^0), and it is a good geometrical exercise to dis-

cover which points P on the circle C give one branch and which
points the other, and whether the asymptotes can be located.
Again, the focal-optical properties of the ellipse and hyperbola
follow more easily from the tangential generations of the curves
than from other methods (**2**).

More can be said about the outlining of a curve by drawing
the tangents to the curve. The parabola is rather special in this
connection. Take two lines ℓ and m (Figure 32^0), and divide each
into equal segments. The scale on the two lines need not be the
same. Number the divisions on each line $1, 2, 3, \ldots$ and join the
points with the same number to each other. If you have avoided
the case of a set of parallel lines, you will obtain the tangents to
a parabola. This also touches ℓ and m.

Carried out with threads of psychedelic colors on a nice piece
of stained wood, these rank as artistic creations and were very
popular (and expensive) some years ago. For more on *curve stitch-
ing* see (**2**).

We began our discussion of the attractions of geometry with
a simple compass, and then we drew straight lines. We continue with the geometry of an unmarked straight ruler, which,
amazingly enough, comprises the whole of projective geometry, of
which Euclidean geometry is a sub-geometry (**1**).

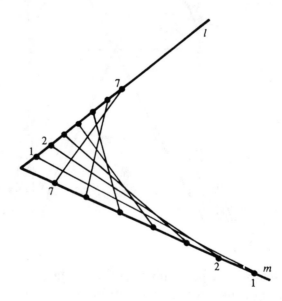

FIG. 32^0

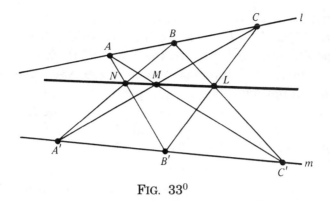

FIG. 33^0

There is a lovely theorem, easily discovered if we doodle with a straightedge, known for over 2,000 years, namely the Pappus Theorem. Take two lines ℓ and m, and mark points A, B, and C on the one, and points A', B', and C' on the other (Figure 33^0). These points can be anywhere on the lines, the theorem being a non-metrical one. Find the intersections $L = BC' \cap B'C$, $M = CA' \cap C'A$, and $N = AB' \cap A'B$.

$$L, M, \text{ and } N \text{ are collinear.}$$

This is surely a startling result, totally unexpected, easy to grasp, and beautiful to contemplate. In the modern foundations of geometry it is one of the fundamental theorems.

If we take other points, say D on ℓ and D' on m, the intersection $CD' \cap C'D$ lies on the line already constructed. This line will not, we note, usually pass through the intersection of ℓ and m.

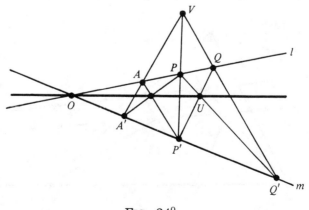

FIG. 34^0

A case where it always does so arises when we choose our points A, A', B, B', and C, C' in a special way, so that the joins AA', BB', and CC' all pass through the same point V.

The theorem of Pappus still applies, and we obtain a figure almost like a national flag, seen in perspective. If $O = \ell \cap m$, we can *prove*, from the construction itself, that the line LMN passes through O. (If we regard O as a point of ℓ, and $O' = O$ as a point of m, where is $AO' \cap A'O$?) This line is called the *polar line* of the point V with regard to ℓ and m.

We now prove what may seem to be, at first sight, a difficult theorem: *If the polar line of V passes through a point U, then the polar line of U passes through the point V.*

Let U be a point on the polar of V, say at the intersection of the lines PQ' and $P'Q$ (Figure 34^0). Then PP' and QQ' both pass through V.

To obtain points on the polar of U with regard to ℓ and m, we must first draw lines through U to intersect ℓ and m. Some already exist, the lines PQ' and QP' meeting ℓ in P and Q and m in Q' and P'.

The intersection of PP' and QQ' gives a point on the polar of U. But this is the point V. We have proved our theorem.

This is certainly very pretty, but is it of any practical use? The answer is in the affirmative ... we can now solve the following problem in civil engineering.

Lines ℓ and m representing roads are drawn, but intersect at an inaccessible point, off the paper (Figure 35^0). It is required

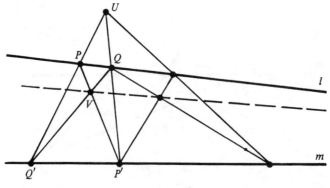

FIG. 35^0

to draw a line through a given point V, between ℓ and m, which will pass through the inaccessible point.

To do this, draw the polar of V with regard to ℓ and m, and mark some point U on this polar, the point U, naturally, being on the paper. Draw the polar of U. It will pass through V, by our theorem, and it also passes through the intersection of ℓ and m.

The notion of a polar line of a point V with regard to two lines can be extended to the idea of a polar line of a point V with regard to a circle, or any conic, to the concept of the polar plane of a point P (as we shall see later in this book) with regard to surfaces of the second degree in space of three dimensions, and in each case harmonic sets of points are involved. Here we use the concept to find the tangents from a point V to a circle (Figure 36^0).

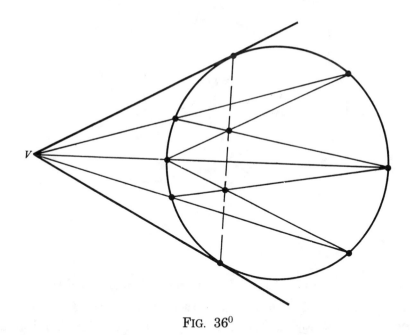

FIG. 36^0

Without deprecating the construction given in Euclid, which involves the use of a compass, it is a matter of experience that those interested in geometry to whom the above construction is new are delighted with its simplicity and beauty.

We end this introduction to the main part of this book with a challenge to the reader. Why does the 5-circle construction given below, with a compass of fixed radius, give the vertex C of an equilateral triangle ABC, given the points A and B? The fixed opening of the compass must exceed half of the distance AB, so as to give intersecting circles. The Euclidean construction uses a compass with opening equal to AB. Some compasses are rusty, and cannot be opened out, hence the above is called a *rusty compass* construction. The 5-circle construction first appeared some years ago as a doodle by a student who was completely unaware of what he had wrought, and has given rise to some very attractive research on rusty compass constructions. There is good reason to believe that the figure has never appeared before (Figure 37^0).

FIG. 37^0

Footnote on the history of geometrical ideas and theorems

While writing this Chapter Zero (March 7, 1994), I was astonished and delighted to hear from Professor Stephen G. Tellman, of the University of Arizona in Tucson, who wrote:

> I vividly remember my reaction of amazement and disbelief when, many years ago, I was first introduced to Feuerbach's Theorem—while browsing through the pages of a small book with the arresting title *Circles,* which I had more or less randomly come upon. That sense of amazement is as alive today as it was then.

Not everyone is interested in the history of geometrical ideas, and in writing a book, too much history may distract the reader from the main discourse. But the story of Karl Wilhelm Feuerbach's tragically short life (1800–1834), as narrated in the warm and sensitive article by Laura Guggenbuhl (3), which Professor Tellman very kindly enclosed with his letter, is one that should be remembered, and it throws light on the history of the ninepoint circle and the great Feuerbach Theorem. See Appendix, p. 89.

Feuerbach's proof of his theorem is analytical and trigono-metrical, using the notion that two circles touch if the distance between their centres is the sum or difference of their radii. He seemed to be unaware of at least three of the points which lie on the ninepoint circle. His paper appeared in 1822 (4). A full discussion of the ninepoint circle as the locus of the centres of rectangular hyperbolas which pass through the vertices of a tri-angle and its orthocentre appeared in a joint work by Brianchon and Poncelet in 1821 (5). Poncelet, who was always very defen-sive about priorities in the discoveries of geometrical theorems (6) never laid claim to the Feuerbach theorem.

For those acquainted with projective geometry, it may be noted that the ninepoint circle appears as a special case of the theorem which discusses the conic locus of the centres of conics which pass through four given points (1).

Finally, it is said that the ninepoint circle is called the Feuer-bach circle in Europe, but it will come as no shock that this does not include England.

References

1. Pedoe, D., *Geometry, A Comprehensive Course,* Dover Publications, 1988.
2. ——, *Geometry and the Visual Arts,* Dover Publications, 1983.
3. Guggenbuhl, Laura, "Karl Wilhelm Feuerbach, Mathematician," *The Scientific Monthly,* vol. 81 (1955) pp. 71–76.
4. Feuerbach, K. W., *Eigenschaften einiger merkwürdiger Punkte des gradlinigen Dreiecks,* Nürnberg, 1822.
5. Brianchon, C. J. and Poncelet, J. V., *Gergonne's Annales de Mathematique,* 11 (1821) p. 205.
6. Pedoe, Dan, "The Principle of Duality," *Mathematics Magazine,* vol. 48, no. 5 (1975) pp. 274–277.

CHAPTER I

THE properties of circles discussed in this chapter are those which
have the habit of appearing in other branches of mathematics.

1. The nine-point circle

This circle is the first really exciting one to appear in any course
on elementary geometry. It is a circle which is found to pass
through nine points intimately connected with a given triangle
ABC. Three proofs are given. Each has its own peculiar merits.

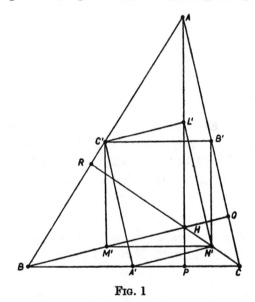

FIG. 1

(i) H is the orthocentre of triangle ABC, the intersection of the
altitudes AP, BQ, CR. The midpoints of BC, CA and AB are
A', B' and C' respectively. L', M', N' are the midpoints of HA,
HB, HC. We show that a circle passes through the nine points
A', B', C', L', M', N', P, Q, R. We use the theorem that the line
joining the midpoints of two sides of a triangle is parallel to the
third side.

Both $B'C'$ and $N'M'$ are $\parallel$ to BC, and both $B'N'$ and $C'M'$ are
$\parallel$ to AH. Hence $B'C'M'N'$ is a rectangle. Similarly $C'A'N'L'$
is a rectangle.

1

Hence $A'L'$, $B'M'$ and $C'N'$ are three diameters of one circle. Since $A'P$ and $L'P$ are $\perp$, this circle passes through P. Similarly it passes through Q and R.

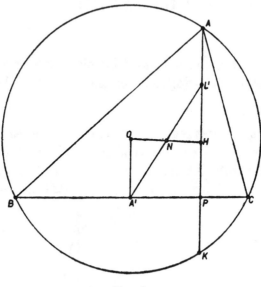

Fig. 2

(ii) Let O be the circumcentre of ABC and R the circumradius. We know that $AH = 2OA'$. Since L' bisects AH, $L'H = OA'$, and therefore OH and $L'A'$ bisect each other, in N, say.

Now $NL' = NA' = NP$ (since $\angle APB = 90°$). We also know that $HP = PK$, and since $ON = NH$, we have

$$NP = \tfrac{1}{2}OK = \tfrac{1}{2}R.$$

Therefore the circle centre N and radius $\tfrac{1}{2}R$ passes through P, A', L'. Similarly it passes through Q, B', M' and R, C', N'.

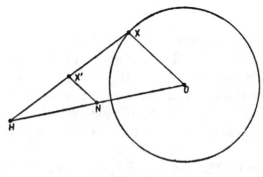

Fig. 3

(iii) The third proof uses a lemma from the theory of similar figures. If X is a point on a given circle of centre O and radius R, and H is any given fixed point, the locus of X', the midpoint of HX, is a circle centre N, where N bisects HO, and radius $\frac{1}{2}R$. The proof is clear.

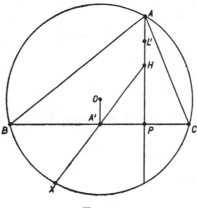

Fig. 4

Applying this lemma to the case of the circumcircle of ABC, H being the orthocentre, we see that a circle centre N and radius $\frac{1}{2}R$ passes through P, Q, R, L', M', N'. We must now show that this circle also passes through A', B', C'. It is sufficient to show that if HX contains A', then HX is bisected at A'.

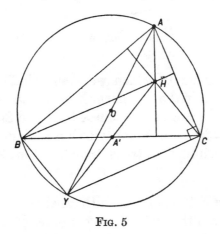

Fig. 5

Let AO meet the circumcircle in Y. Then $\angle ACY = 90°$, so that $YC \parallel BH$. Since also $\angle ABY = 90°$, $YB \parallel CH$. Therefore $YBHC$ is a parallelogram, and the diagonals YH and BC

bisect each other. Hence Y is the same point as X in the previous diagram, and the theorem is proved.

The methods used above are elementary, and at least one of the three proofs will be familiar to most readers of this book. In chapter II we shall need to assume some theorems in inversion and coaxal systems of circles. As fashions in mathematical teaching change, it cannot be assumed that all readers will be familiar with these theories, and a brief (but adequate) outline of each will now be given.

2. Inversion

This is a one-to-one transformation of the points of the plane by

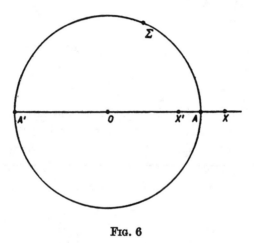

FIG. 6

means of a given circle, which we shall call Σ, of radius k and centre O. To obtain the transform X' of any given point X, or the *inverse* of X in the circle Σ, we join X to O, and find X' on OX such that $OX.OX' = k^2$. The point O itself is excluded from the points of the plane which may be transformed. It is clear that

    (i) X is the inverse of X';
    (ii) points on Σ transform into themselves;
    (iii) if A, A' are the ends of the diameter of Σ through X, then X, X' are harmonic conjugates with respect to A, A'.

All circles through X and X' cut Σ orthogonally, since the square of the tangent from O to such a circle $= OX.OX' = k^2$.

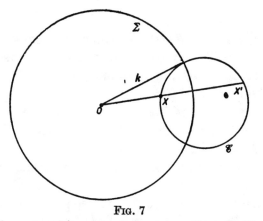

FIG. 7

As a corollary to this remark, we see that if $\mathscr{C}$ is any circle orthogonal to Σ, all points on $\mathscr{C}$ invert into points on $\mathscr{C}$, for $OX.OX' = k^2$. It is sometimes necessary to consider the process of inversion when Σ, the circle of inversion, is a line. If we keep A fixed, and let A' move off to infinity, we see that X' moves towards the geometrical image of X in the resulting line.

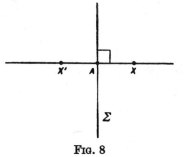

FIG. 8

For future application we now give a construction, using only a

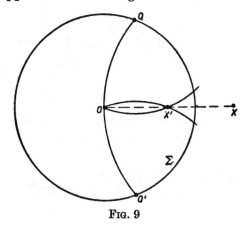

FIG. 9

pair of compasses, by which we can find the inverse of a point X in a circle Σ of given centre O. Let the circle centre X and radius XO cut Σ in Q and Q', and let O and X' be the intersections of the two circles with centres Q and Q' and radii QO and $Q'O$ respectively. Then X' is the inverse of X in Σ.

For QOX and $X'OQ$ are isosceles triangles with a common base angle. They are therefore similar, and $OX : OQ = OQ : OX'$. Hence $OX.OX' = OQ^2$.

This construction lies at the basis of the discussion on *compass geometry* in §11.

We are interested in the locus of inverse points X' as X describes a given curve $\mathscr{C}$. This locus is called the *inverse of the curve* $\mathscr{C}$ with respect to Σ. When $\mathscr{C}$ is a circle orthogonal to Σ, we have seen that the inverse is $\mathscr{C}$ itself. The general theorem is:

The inverse of a circle is a straight line or a circle.

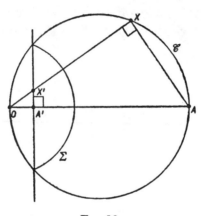

FIG. 10

(i) The centre of inversion O is on $\mathscr{C}$:

Let OA be the diameter of $\mathscr{C}$ through O, and let A' be the inverse of A. Then $OA.OA' = k^2$. If X is any point on $\mathscr{C}$, and X' its inverse,

$$OX.OX' = k^2 = OA.OA'.$$

Hence A, X, X', A' are concyclic points, and since $\angle AXO = 90°$, $\angle X'A'A = 90°$. Therefore X' describes the line through $A' \perp$ to OA.

We note that this line is $\parallel$ to the tangent to $\mathscr{C}$ at O. It naturally passes through the intersections of $\mathscr{C}$ and Σ, since points on Σ invert into themselves.

Corollary. The inverse of a straight line is a circle through the centre of inversion.

(ii) The centre of inversion O is not on $\mathscr{C}$. Let C be the centre

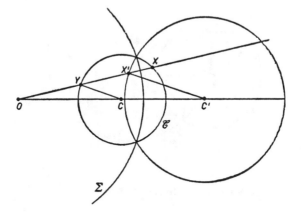

FIG. 11

of $\mathscr{C}$. Since

$$OX \cdot OX' = k^2,$$

and

$$OX \cdot OY = t^2,$$

where t is the tangent[†] from O to $\mathscr{C}$, by division we have

$$\frac{OX'}{OY} = \frac{k^2}{t^2}.$$

Since this ratio is constant, it follows that if C' is a point on OC such that $C'X' \parallel CY$, then C' is a fixed point, and $C'X'$ is constant. Hence X' moves on a circle centre C'.

Note that C' is not necessarily the inverse of C. We shall soon find what point the centre of $\mathscr{C}$ does invert into.

We need only one further theorem from the theory of inversion:

The angle of intersection of two circles is unaltered by inversion.

In fact a more general theorem is true, but this one will suffice for our subsequent needs.

---

† We shall see later that t^2 may be negative. Then O is inside $\mathscr{C}$.

(i) If both circles $\mathscr{C}$ and $\mathscr{D}$ pass through the centre O of inversion, they transform into straight lines parallel to the tangents to

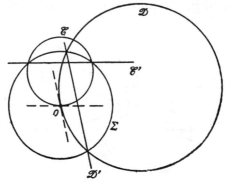

Fig. 12

$\mathscr{C}$ and $\mathscr{D}$ at O, so that the theorem is evident.

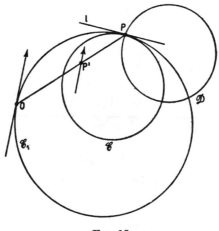

Fig. 13

(ii) If at least one of the circles $\mathscr{C}$, $\mathscr{D}$ does not pass through O, let P be a point of intersection of the two circles, and let l, m be the tangents at P to $\mathscr{C}$, $\mathscr{D}$ respectively. We may draw a circle $\mathscr{C}_1$ through O which touches l at P, and a circle $\mathscr{D}_1$ through O which touches m at P. Then the angle between $\mathscr{C}_1$ and $\mathscr{D}_1$ is the same as that between $\mathscr{C}$ and $\mathscr{D}$.

Now if P' is the inverse of P, the tangent at P' to the inverse of $\mathscr{C}$ is the inverse of $\mathscr{C}_1$. Hence the angle between the circles inverse to $\mathscr{C}$ and to $\mathscr{D}$ is equal to the angle between $\mathscr{C}_1$ and $\mathscr{D}_1$ (by (i) above), and this is equal to the angle between $\mathscr{C}$ and $\mathscr{D}$.

With the help of this theorem we can now find the inverse of the centre of a circle $\mathscr{C}$. All lines through the centre C cut $\mathscr{C}$ orthogonally. These lines invert into circles through the centre of inversion O and the inverse of C. Since all circles through O and the inverse of C cut the inverse of $\mathscr{C}$ orthogonally, the inverse of C must be *the inverse of O in the inverse of $\mathscr{C}$*.

If O lies on $\mathscr{C}$, so that the inverse of $\mathscr{C}$ is a straight line, the centre of $\mathscr{C}$ inverts into the geometrical image of O in this line.

We now give some applications of the theory of inversion.

3. Feuerbach's theorem

The nine-point circle of a triangle touches the incircle and excircles of the triangle.

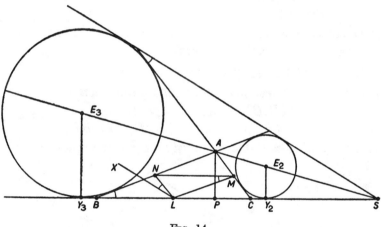

Fig. 14

We consider the excircles opposite B and C. Let their centres be E_2 and E_3, and their points of contact with BC Y_2 and Y_3. The points A, S divide E_2E_3 internally and externally in the same ratio, that of the radii of the two excircles. Dropping perpendiculars on BC, it follows that P and S also divide Y_2Y_3 internally and externally in the same ratio. That is, P and S are harmonic conjugates with respect to Y_2 and Y_3.

Let L, M, N be midpoints of BC, CA, AB. We note that $BY_2 = \frac{1}{2}(BC + CA + AB) = CY_3$, so that L is also the midpoint of Y_2Y_3. From the harmonic relation just proved we see that

$$LP . LS = LY_2{}^2 = LY_3{}^2.$$

Now let LX be the tangent at L to the circle LMN, where X is on the same side of BC as N. Then
$$\angle XLN = \angle NML = \angle ABC.$$
Hence the angle between LX and $CA = \angle ABC$. It follows that LX is $\parallel$ to the fourth common tangent of the two excircles, the common tangent through S other than BC.

Now since $LP.LS = LY_2{}^2 = LY_3{}^2$, the nine-point circle LMN is the inverse of this fourth common tangent with respect to the circle centre L and radius LY_2. For this nine-point circle contains P, and inverts into the line through S parallel to the tangent to the circle at L.

Also since the circle of inversion cuts both excircles orthogonally, each excircle inverts into itself. We have seen that the inverse of the nine-point circle touches these excircles. It follows that the nine-point circle touches the excircles centres E_2, E_3. In precisely the same way we can show that the nine-point circle touches the incircle and the excircle opposite A.

4. Extension of Ptolemy's theorem

If A, B, C, D are any four points in a plane, then
$$AB.CD + AD.BC > AC.BD,$$
unless A, B, C, D lie, in the order $ABCD$, on a circle or a straight line. In the latter case, the inequality becomes an equality.

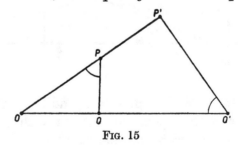

FIG. 15

This is proved by noting the influence of inversion on the length of a segment. If the points P, Q invert into the points P', Q', with O as centre of inversion, then OPQ and $OQ'P'$ are similar triangles. Hence
$$\frac{P'Q'}{PQ} = \frac{OP'}{OQ} = \frac{OP.OP'}{OP.OQ} = \frac{k^2}{OP.OQ},$$
so that
$$P'Q' = \frac{k^2\,PQ}{OP.OQ}.$$

Now, given the four points A, B, C, D, we invert with respect to A. Let B', C', D' be the respective inverses of B, C, D. Then $B'C' + C'D' > B'D'$ unless C' lies on the line $B'D'$ between B' and D'. In the latter case we have

$$B'C' + C'D' = B'D'.$$

The first inequality becomes

$$\frac{BC}{AB.AC} + \frac{CD}{AC.AD} > \frac{BD}{AB.AD},$$

or

$$AB.CD + AD.BC > AC.BD,$$

unless C' lies on the line $B'D'$ between B' and D', in which case A, B, C, D lie on a circle in the order $ABCD$, or on a straight line.

5. Fermat's problem

A, B, C are any three points in a plane. *To find a point P such that $PA + PB + PC$ shall be least.*

Let B and C be the acute angles of the triangle ABC. On BC, and away from A, describe an equilateral triangle BCD. Then by the extension of Ptolemy's theorem, unless P lies on the circle

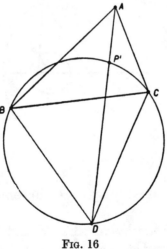

Fig. 16

BCD so that the order of the points is $BPCD$, we have

$$BP.CD + PC.DB > PD.BC,$$

or

$$PB + PC > PD,$$

since $CD = DB = BC$. Therefore

$$PA + PB + PC > PA + PD.$$

Now, unless P lies on AD, we have $PA + PD > AD$. Hence, unless P is at P' (the other intersection of AD with the circle BCD), we have $PA + PB + PC > AD$.

But if P is at P', both the above inequalities become equalities; so that $P'A + P'B + P'C = AD$.

Hence $P'A + P'B + P'C < PA + PB + PC$.

Therefore P' is the required point.

If $\angle BAC = 120°$, $A = P'$, and A is the required point.

If $\angle BAC > 120°$, A is still the required point.

6. The centres of similitude of two circles

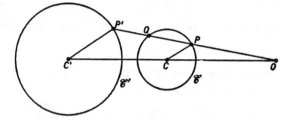

FIG. 17

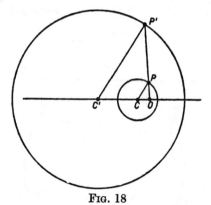

FIG. 18

These points have already appeared in some of our proofs. We now define them.

External centre of similitude. Let $\mathscr{C}$ and $\mathscr{C}'$ be two circles of unequal radii, with centres C and C'. Let CP and $C'P'$ be parallel radii drawn in the *same* sense. Then it is clear that $P'P$ cuts $C'C$ in a fixed point O, which divides $C'C$ externally in the ratio of the radii of the two circles. This is the external centre of similitude. If the ratio of the radii $= k$, then $OP' : OP = k$. We also have $OP.OQ = $ constant. Therefore

$$OP'.OQ = \text{constant}.$$

It follows that with centre of inversion O, and a suitable radius of inversion, $\mathscr{C}$ can be inverted into $\mathscr{C}'$.

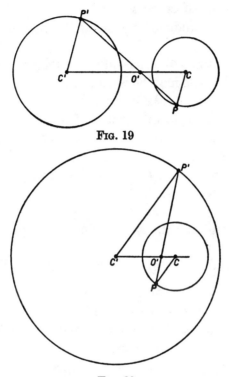

FIG. 19

FIG. 20

Internal centre of similitude. Here the radii of the circles $\mathscr{C}$ and $\mathscr{C}'$ may be equal. CP and $C'P'$ are parallel radii in opposite senses. $P'P$ cuts $C'C$ in a fixed point O', the internal centre of similitude, which divides $C'C$ internally in the ratio of the radii.

Direct common tangents of $\mathscr{C}$ and $\mathscr{C}'$, if they exist, cut the line of centres in O, and transverse common tangents, if they exist, cut the line of centres in O'.

If $\mathscr{C}$ and $\mathscr{C}'$ are concentric, the common centre is both the internal and external centre of similitude.

If $\mathscr{C}$ and $\mathscr{C}'$ touch externally, that is, if they lie on opposite sides of the tangent at the point of contact, this point is the *internal* centre of similitude. If the circles touch internally, the point of contact is the *external* centre.

If we have three circles, no two of which have equal radii, and the centres of which form a triangle, the six centres of similitude lie in threes on four straight lines. This is easily proved, using the theorem of Menelaus.

7. Coaxal systems of circles

These are best introduced by using the methods of coordinate geometry.

The equation of any line passing through the origin of coordinates in the (x,y)-plane may be written as
$$\lambda x + \mu y = 0.$$
These lines are said to form a *pencil* of lines, and the equation of the system depends *linearly* on two distinct lines of the pencil, given by $x = 0$ and $y = 0$.

Similarly, if $u \equiv ax + by + c = 0$ and $v \equiv a'x + b'y + c' = 0$ are any two distinct lines, the line
$$\lambda u + \mu v = 0$$
represents, for suitable λ, μ, any line through the intersection of $u = 0$ and $v = 0$.

When we consider two distinct circles $\mathscr{C}$ and $\mathscr{C}'$, given by the equations
$$C \equiv x^2 + y^2 + 2gx + 2fy + c = 0,$$
$$C' \equiv x^2 + y^2 + 2g'x + 2f'y + c' = 0,$$
the system $\lambda C + \mu C' = 0$, derived linearly from them, is a system of circles, since it may also be written as
$$x^2 + y^2 + \frac{2x(\lambda g + \mu g')}{\lambda + \mu} + \frac{2y(\lambda f + \mu f')}{\lambda + \mu} + \frac{\lambda c + \mu c'}{\lambda + \mu} = 0.$$
If (x',y') is a point of intersection of $\mathscr{C}$ and $\mathscr{C}'$,
$$C(x',y') = C'(x',y') = 0,$$
that is, the point satisfies the equations of both circles. Since also, for all values of λ, μ,
$$\lambda C(x',y') + \mu C'(x',y') = 0,$$
this same point lies on all circles of the system $\lambda C + \mu C' = 0$. Such a system of circles therefore passes through the points of intersection (if any) of $\mathscr{C}$ and $\mathscr{C}'$. The system may be described as a *pencil* of circles, but for reasons which will soon become clear, we give it the more normal title of a *coaxal system* of circles.

A geometrical meaning may be given to the ratio of the parameters $\lambda : \mu$. Let $P(x',y')$ be any point in the plane, and let any line through P cut $\mathscr{C}$ in Q and R. Then it is elementary that the product
$$PQ . PR = x'^2 + y'^2 + 2gx' + 2fy' + c.$$

This constant, which is > 0 if P is outside the circle and equal to the square of the tangent from P to $\mathscr{C}$, is called, in all cases, the *power* of P with respect to $\mathscr{C}$. If P is inside $\mathscr{C}$, the power is negative, and it is zero, of course, if P is on $\mathscr{C}$. Since

$$\lambda C + \mu C' = 0$$

may be written as

$$\frac{\lambda}{\mu} = -\frac{C'}{C},$$

we see that *the locus of a point which moves so that the ratio of its powers with respect to two given circles $\mathscr{C}$ and $\mathscr{C}'$ is a constant is a circle of the coaxal system determined by $\mathscr{C}$ and $\mathscr{C}'$.*

A particular and important member of the system is obtained by taking the ratio of the powers to be unity. We then obtain

$$C - C' \equiv 2(g - g')x + 2(f - f')y + c - c' = 0,$$

a straight line, called the *radical axis* of the system. If $\mathscr{C}$ and $\mathscr{C}'$ intersect, this line is the common chord of the two circles, and it is evident that points on the common chord of two circles do have equal powers with respect to each circle. But the radical axis

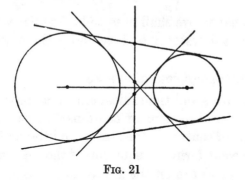

FIG. 21

exists even if the circles do not intersect. It is easily verified that the radical axis of two circles is perpendicular to the line of centres.

We deduce that the midpoints of the common tangents of two circles lie on a line which is perpendicular to the line of centres.

We obtain the same coaxal system of circles if we take any two distinct members of the system $\lambda C + \mu C' = 0$, say

$$C_1 \equiv \lambda_1 C + \mu_1 C' = 0,$$

and

$$C_2 \equiv \lambda_2 C + \mu_2 C' = 0,$$

where $\lambda_1/\mu_1 \neq \lambda_2/\mu_2$, and write down the system of circles which depend linearly on C_1 and C_2. This should be verified. Another theorem of importance is that any two circles of a coaxal system have the same radical axis. This explains the term *coaxal system*. We can verify this directly. The radical axis of C_1 and C_2 is

$$2x \left[\frac{\lambda_1 g + \mu_1 g'}{\lambda_1 + \mu_1} - \frac{\lambda_2 g + \mu_2 g'}{\lambda_2 + \mu_2} \right] + 2y \left[\frac{\lambda_1 f + \mu_1 f'}{\lambda_1 + \mu_1} - \frac{\lambda_2 f + \mu_2 f'}{\lambda_2 + \mu_2} \right]$$

$$+ \left[\frac{\lambda_1 c + \mu_1 c'}{\lambda_1 + \mu_1} - \frac{\lambda_2 c + \mu_2 c'}{\lambda_2 + \mu_2} \right] = 0.$$

This easily reduces to

$$(\lambda_1 \mu_2 - \lambda_2 \mu_1)[2x(g - g') + 2y(f - f') + c - c'] = 0.$$

It follows that the centres of all circles of a coaxal system lie on a line perpendicular to the radical axis. This is also clear from the fact that the centre of $\lambda C + \mu C' = 0$ is the point

$$\left[\frac{\lambda(-g) + \mu(-g')}{\lambda + \mu}, \quad \frac{\lambda(-f) + \mu(-f')}{\lambda + \mu} \right].$$

In the next chapter we shall need to be familiar with a few more properties of coaxal systems, and we derive them here.

8. Canonical form for coaxal system

The line of centres and the radical axis of a coaxal system offer, in a sense, a natural system of coordinate axes for the system. We take the line of centres as the x-axis, and the radical axis as the y-axis, and see what form the equation of the system takes.

Since their centres are on the x-axis, any two circles $\mathscr{C}$ and $\mathscr{C}'$ have equations of the form,

$$C \equiv x^2 + y^2 + 2gx + c = 0,$$

and

$$C' \equiv x^2 + y^2 + 2g'x + c' = 0.$$

The radical axis of these two circles is

$$C - C' \equiv 2(g - g')x + c - c' = 0,$$

and since this is to be $x = 0$, we must have $c = c'$. The circles of the system may therefore be written in the form

$$x^2 + y^2 + 2\lambda x + c = 0,$$

where c is constant, and λ a parameter. This is a canonical, or standard form for a coaxal system.

We note that the system now depends linearly on the circle $x^2 + y^2 + c = 0$, and the radical axis $x = 0$.

We can now easily distinguish between the intersecting and non-intersecting type of coaxal system. Intersections, if any, must lie on the radical axis $x = 0$. This meets circles of the system where $y^2 + c = 0$. If c is positive, there are no intersections, and we have the non-intersecting type of coaxal system. If c is negative, all circles of the system cut the radical axis in the same two points.

(i) $c = k^2 > 0$. Any circle of the system may be written in the form

$$(x + \lambda)^2 + y^2 = \lambda^2 - k^2.$$

For real circles, we must have $\lambda^2 \geqslant k^2$. There are two circles of zero radius in the system, with centres at $(-k, 0)$ and $(k, 0)$. These points, which we denote by L and L', are called the *limiting*

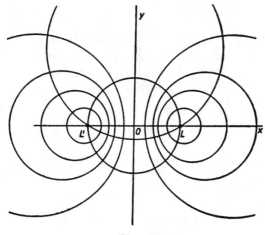

Fig. 22

points of the system. No circles of the system have their centres between L and L'. If a circle of the system cuts the line of centres in A, A', then $OA \cdot OA' = k^2 = OL^2$, and it follows that L and L' are harmonic conjugates with respect to A and A'. Hence any circle through L and L' cuts all circles of the coaxal system orthogonally.

(ii) $c = -k^2 < 0$. This case presents no features which are

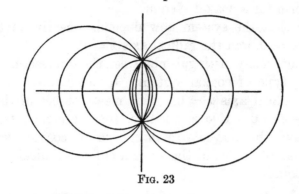

Fig. 23

not evident geometrically.

(iii) $c = 0$. This can be regarded as an intermediate case.

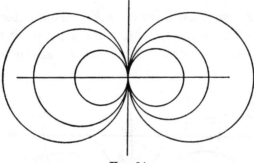

Fig. 24

All circles of the system touch the radical axis, and the point of contact may be regarded as consisting of two coincident limiting points.

We have remarked that in (i) any circle through L and L' is orthogonal to every circle of the given system. The equation of a circle through L and L' is

$$x^2 + y^2 + 2\mu y - k^2 = 0,$$

and this, of course, is a coaxal system of the intersecting type. The orthogonality relation is evident algebraically, since the condition to be satisfied, $2gg' + 2ff' - c - c' = 0$, becomes here $2\lambda.0 + 2.0.\mu - k^2 + k^2 = 0$, which is the case.

Hence, associated with every coaxal system of a given type there is a coaxal system of the opposite type, and the circles of one cut the circles of the other orthogonally. Such systems are called *conjugate systems,* for reasons which will become clear in chapter II.

9. Further properties

If we try to find the circles of the coaxal system
$$x^2 + y^2 + 2\lambda x + c = 0$$
which pass through $P(x',y')$, we see that

$$\lambda = -\frac{(x'^2 + y'^2 + c)}{2x'} ;$$

that is, there is a unique value of the parameter which determines the circle. Hence one circle of a coaxal system passes through a given point P. The construction of this circle is evident for intersecting systems, since a circle is determined by three points, but not so evident for a non-intersecting system. In this case we invoke the *conjugate system*, which is of the intersecting type, and draw the unique circle of this system which passes through P. We can then construct the required circle by drawing the unique circle through P which is orthogonal to the constructed circle and has its centre on the line of centres of the given coaxal system.

If we invert a coaxal system of circles, it is fairly clear that we obtain an intersecting coaxal system in the intersecting case, but this is not so clear in the non-intersecting case. However we can bring in the *conjugate system*, which is of the intersecting type, and remembering that circles which are orthogonal invert into orthogonal circles, we see that the circles of the non-intersecting system invert into circles which are orthogonal to all the circles of a coaxal system. Such a system of circles must form the conjugate coaxal system.

However, inversion is a transformation which sometimes does more than we demand of it, and the reader may feel reluctant to accept the proof indicated when we show, as we do now, that *any non-intersecting system of coaxal circles may be inverted into a system of concentric circles*. The question he will ask is: if a system of concentric circles is a coaxal system, why have we not come across it in the above treatment?

We leave the reader to ponder on this apparent paradox, and prove the theorem. Let L and L' be the limiting points of the non-intersecting coaxal system, and invert with respect to a circle centre L. Then the circles of the system invert into circles which are orthogonal to the inverses of all circles through L and L'. But these inverses are straight lines through a fixed point, the inverse of L'. Hence the coaxal system inverts into a system of concentric circles, with centre at the inverse of L'.

This theorem provides a solution of the attractive problem:

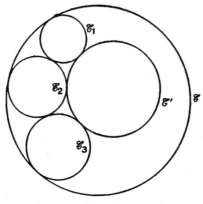

FIG. 25

given two circles $\mathscr{C}$ and $\mathscr{C}'$, one lying inside the other, can we find a chain of circles, $\mathscr{C}_1$, $\mathscr{C}_2$, $\mathscr{C}_3$, ..., $\mathscr{C}_n$, all of which touch $\mathscr{C}$ and $\mathscr{C}'$, such that $\mathscr{C}_1$ touches $\mathscr{C}_2$ and $\mathscr{C}_n$, $\mathscr{C}_2$ touches $\mathscr{C}_3$ and $\mathscr{C}_1$, ..., $\mathscr{C}_n$ touches $\mathscr{C}_{n-1}$ and $\mathscr{C}_1$?

Inverting $\mathscr{C}$ and $\mathscr{C}'$ into concentric circles, we see that the

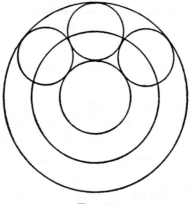

FIG. 26

problem does not always have a solution, but that when there is a solution, we can begin the chain by drawing any circle which touches $\mathscr{C}$ and $\mathscr{C}'$, and can then fit in the other members of the chain in an obvious manner.

We conclude this discussion of coaxal circles by considering the radical axes, in pairs, of three circles $\mathscr{C}$, $\mathscr{C}'$, $\mathscr{C}''$. The intersection of two of these radical axes is a point which has equal powers with respect to all three circles.   This point is therefore on the third radical axis.   Hence, if they are not parallel, the radical axes of the three pairs of circles which can be formed from three given circles are concurrent.   The point of concurrence is called the *radical centre* of the three circles.   The circle which has its centre at the radical centre, and radius equal to the length of the tangent from the centre to any one of the three circles, if it lies outside one, and therefore all three circles, cuts each of the three circles orthogonally and is the only circle which does this.   With respect to this circle as circle of inversion, each of the three circles inverts into itself.

10. Problem of Apollonius

Given three circles $\mathscr{C}_1$, $\mathscr{C}_2$ and $\mathscr{C}_3$, to draw a circle touching all three.

We need two preliminary results:

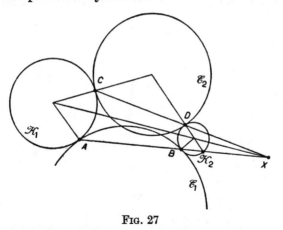

Fig. 27

(i) If two non-concentric circles $\mathscr{K}_1$, $\mathscr{K}_2$ touch two other non-concentric circles $\mathscr{C}_1$, $\mathscr{C}_2$ at A, B and C, D respectively, and if the contacts at C and D are of like[†] or unlike type according as those at A and B are, then AB and CD meet in a centre of similitude of the circles $\mathscr{K}_1$, $\mathscr{K}_2$, or they are parallel to the line of centres of these circles.

To prove this result we join the centres of $\mathscr{K}_1$, $\mathscr{K}_2$ and $\mathscr{C}_1$ and find, using the theorem of Menelaus, where AB cuts the line joining the centres of $\mathscr{K}_1$, $\mathscr{K}_2$.   If the given conditions are satisfied, and we consider the triangle formed by the centres of $\mathscr{K}_1$,

† Contacts are of " like type " when both are external or both are internal.

$\mathcal{K}_2$ and $\mathscr{C}_2$, we see that CD passes through the same point X, and that this is a centre of similitude of $\mathcal{K}_1$ and $\mathcal{K}_2$.

(ii) We remarked earlier (see page 12) that $XA \cdot XB = XC \cdot XD$, so that X is on the radical axis of $\mathscr{C}_1$ and $\mathscr{C}_2$.

We also note that if the contacts at A and B are of like or unlike type according as those at C and D are, then the contacts at A and C are of like or unlike type according as those at B and D are. Therefore AC and BD meet at a centre of similitude of $\mathscr{C}_1$ and $\mathscr{C}_2$, and this centre is on the radical axis of $\mathcal{K}_1, \mathcal{K}_2$.

We have therefore shown that the radical axis of $\mathscr{C}_1$ and $\mathscr{C}_2$ contains a centre of similitude of $\mathcal{K}_1$ and $\mathcal{K}_2$, and the radical axis of $\mathcal{K}_1$ and $\mathcal{K}_2$ contains a centre of similitude of $\mathscr{C}_1$ and $\mathscr{C}_2$.

We now return to our problem, and assume, for simplicity, that the radii of $\mathscr{C}_1$, $\mathscr{C}_2$, $\mathscr{C}_3$ are three distinct numbers, so that all the centres of similitude exist.

Let $\mathcal{K}_0$ be a circle touching the three circles externally. Inver-

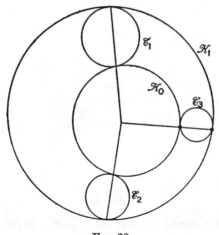

Fig. 28

sion from the radical centre of $\mathscr{C}_1$, $\mathscr{C}_2$, $\mathscr{C}_3$ changes these circles into themselves, and $\mathcal{K}_0$ is transformed into a circle $\mathcal{K}_1$ which is touched internally by the three circles. The joins of the points of contact of each circle $\mathscr{C}_i$ ($i = 1, 2, 3$) with $\mathcal{K}_0$ and with $\mathcal{K}_1$ pass through the radical centre.

The pole, for $\mathscr{C}_1$, of the line joining the points of contact of $\mathcal{K}_0$ and $\mathscr{C}_1$ and of $\mathcal{K}_1$ and $\mathscr{C}_1$ is the intersection of the tangents at these points of contact. These tangents are the radical axes of $\mathscr{K}_0$ and $\mathscr{C}_1$ and of $\mathscr{K}_1$ and $\mathscr{C}_1$, and since the radical axes of three circles, taken in pairs, are concurrent, this pole lies on the radical axis of $\mathscr{K}_0$ and $\mathscr{K}_1$.

By the reciprocal property of pole and polar, the polar of this point contains the pole, for $\mathscr{C}_1$, of the radical axis. Now the polar is the line joining the points of contact of $\mathscr{K}_0$ and $\mathscr{C}_1$ and of $\mathscr{K}_1$ and $\mathscr{C}_1$ and passes through the radical centre of $\mathscr{C}_1, \mathscr{C}_2, \mathscr{C}_3$. Since also the radical axis of $\mathscr{K}_0$ and $\mathscr{K}_1$ contains the external centres of similitude of the circles $\mathscr{C}_1, \mathscr{C}_2, \mathscr{C}_3$ taken in pairs (this was proved in (ii) above), we have the following construction, due to Gergonne:

Draw the line through the external centres of similitude of $\mathscr{C}_1, \mathscr{C}_2$ and $\mathscr{C}_3$, taken in pairs, and join its poles for each of these circles to the radical centre of the three circles. These joins cut the given circles in points of contact of two of the required circles.

The centres of similitude of $\mathscr{C}_1, \mathscr{C}_2, \mathscr{C}_3$ lie by threes on four lines, and each line of centres of similitude gives two tangent circles. In the most favourable case we therefore expect to find *eight* solutions of the problem of Apollonius.

We shall verify this from another point of view in chapter II.

11. Compass geometry

We conclude this chapter by showing how the straight edge may be dispensed with in the elementary constructions of Euclidean geometry.

A straight edge is used to connect two given points by a straight line. No pair of compasses will do this. At the same time the straight edge and compasses are primarily used to determine certain points. The fundamental problems in Euclidean geometry are:

(i) to find the intersections of two circles, the centre and radius of each being given;

(ii) to find the intersection of a line given by two points with a circle given by its centre and radius;

(iii) to find the intersections of two lines, each given by two points.

The solution of (i) is naturally possible using only the compasses. Problems (ii) and (iii) reduce to (i) after inversion.

Therefore, to show that all Euclidean constructions are possible using only a pair of compasses, we must show:

    (a) how to find the inverse of a given point in a circle of given centre and radius;

    (b) how to find the centre of a circle which is to pass through three given points.

We gave the solution of (a) in §2, page 5.

Before dealing with (b), we show how to find, using compasses only, the midpoint of the segment determined by two given points. In the first place we show how to extend a given segment AO to

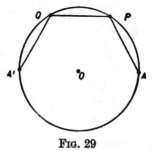

Fig. 29

twice its length. Draw the circle centre O and radius OA, and mark off the chords $AP = PQ = QA' = OA$. Then AOA' is a straight line, and $AA' = 2OA$.

We can now find the midpoint of the segment OA. Extend AO to A', where $AO = OA'$, and find the inverse of A' in the

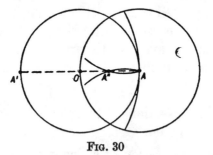

Fig. 30

circle centre A and radius AO. If A'' is the inverse,
$$AA''.AA' = AO^2,$$
and since $AA' = 2AO$, the point A'' is the midpoint of OA.

We can extend OA to A', where $OA' = nOA$, n being any positive integer, and the inverse of A' in the circle centre A and radius AO gives OA/n.

(b) Let A, B, C be the three given points. Invert with respect to the circle centre A and radius AB. The circle ABC inverts

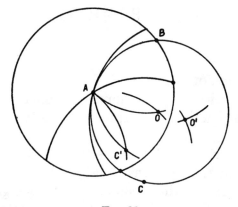

FIG. 31

into the line BC', where C' is the inverse of C in the circle of inversion. This point can be constructed. The centre of circle ABC inverts into the inverse of the centre of inversion, which is A, in the inverse of circle ABC, namely, the line BC' (see §2, page 9). The inverse of A in BC' is the geometrical image of A in BC'. This is easily constructed, as indicated in Fig. 31. If this geometrical image be O', we finally find the inverse of O' in the circle of inversion. This final point O is the centre of circle ABC.

The reader should carry out this construction for himself.

To find where two lines determined by points A,B and C,D respectively meet, we invert with respect to any suitable circle, centre O, say. If the respective inverse points are A',B' and C',D', the intersection of the circles $OA'B'$ and $OC'D'$, other than O, is the inverse of the point required. We can draw these circles, by the above construction, and find this point.

Bright students often rediscover this compass geometry for themselves, and are dashed to find that it was fully investigated, without the benefit of inversion, by Mascheroni in *La Geometria del Compasso*, Pavia, 1797.

CHAPTER II

1. Representation of a circle

In this chapter we adopt a completely different point of view from that pursued in the first chapter. We note that the equation of a circle is

$$C \equiv x^2 + y^2 + 2gx + 2fy + c = 0, \tag{1}$$

and that this equation contains three parameters (g, f, c). If we are given the values of these parameters, we are given the circle $\mathscr{C}$ represented by the equation $C = 0$. Conversely, the equation of a circle $\mathscr{C}$ can be written in only one way in the form (1). Hence, any circle can be *represented* by the three numbers (g, f, c).

It is natural to think of three numbers (g, f, c) as giving the three coordinates of a point P in three-dimensional space. We shall amplify this statement in a moment. But assuming that we understand what is implied by the term "three-dimensional space", we can now say that circles in the (x,y)-plane can be *represented* by points in three-dimensional space (x, y, z). In this chapter we shall study the properties of this representation.

2. Euclidean three-space, E_3

Let Ox, Oy, Oz be three mutually perpendicular straight lines drawn through a point O. If P is any point, we consider the

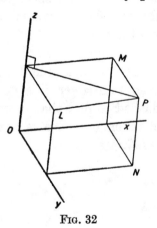

Fig. 32

perpendiculars PL, PM and PN drawn from P to the planes Oyz, Ozx and Oxy respectively. The distance LP measures the

26

x-coordinate of P, and is positive if LP is in the direction Ox, negative if it is in the opposite direction. Similarly for the y- and z-coordinates of P. If the coordinates of P are (x_1,y_1,z_1), P is the intersection of the three planes $x = x_1$, $y = y_1$, and $z = z_1$.

We do not intend to develop the coordinate geometry of E_3, but a few results are essential for our purpose. The perpendicular from P (x_1,y_1,z_1) on to the line Oz is easily seen to be $\sqrt{(x_1{}^2 + y_1{}^2)}$.

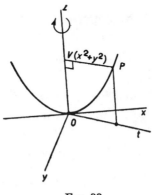

FIG. 33

If we imagine the parabola in the (z,t) plane whose equation is $z - t^2 = 0$, this parabola touches the t-axis at the origin. If we rotate the parabola about the z-axis, we obtain a *paraboloid of revolution*. Since the perpendicular from any point P (x,y,z) of the paraboloid on to the z-axis $= \sqrt{(x^2 + y^2)} = t = \sqrt{z}$, we have
$$x^2 + y^2 - z = 0$$
for the equation of the paraboloid. This surface will play an important part in our investigations.

The only other fact we shall need is that the Joachimstahl ratio-formulae continue to hold in E_3. In other words, if P (x,y,z) and Q (x',y',z') are two points in E_3, the point which divides PQ in the ratio $\lambda : 1$ has coordinates
$$\left(\frac{x + \lambda x'}{1 + \lambda}, \ \frac{y + \lambda y'}{1 + \lambda}, \ \frac{z + \lambda z'}{1 + \lambda} \right).$$

This is derived immediately from the formula in the two-dimensional case by projecting P, Q and the point whose coordinates are required on to the plane Oxy, say.

We now return to the representation of a circle $\mathscr{C}$. It is more convenient to take its equation in the form

$$C \equiv x^2 + y^2 - 2\xi x - 2\eta y + \zeta = 0 \qquad (1)$$

rather than that given by Eq. (1), §1. For the centre of $\mathscr{C}$ is (ξ,η), and if we suppose that $\mathscr{C}$ lies in the plane Oxy of E_3, and represent $\mathscr{C}$ by the point P (ξ,η,ζ), *the orthogonal projection of P on to the plane Oxy is the centre of the circle $\mathscr{C}$ which is represented by the point P.*

3. First properties of the representation

The square of the radius of the circle $\mathscr{C}$ given by (1) is $\xi^2 + \eta^2 - \zeta$. Circles of zero radius are represented by points P (x,y,z) which satisfy the equation

$$\Omega \equiv x^2 + y^2 - z = 0. \qquad (1)$$

As we saw in §2, this equation represents a paraboloid of revolution. This paraboloid Ω plays a fundamental part in our investigation. If, at the point $(x,y,0)$, we erect a perpendicular to the

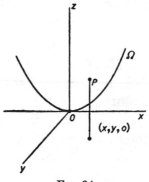

FIG. 34

plane Oxy, this meets Ω in a unique point P, which represents the circle centre $(x,y,0)$ and of zero radius. Points above P (in an obvious sense) have a z-coordinate which exceeds that of P, and therefore represent circles the square of whose radius is negative. We call such circles *imaginary* circles, noting that the centre of such a circle is a real point. Points below P represent *real* circles, or the ordinary circles of geometry. Since points above P lie inside Ω, and points below P lie outside Ω, we see that *the paraboloid separates the points which represent real circles from those which represent imaginary circles.*

The preceding remarks also show that circles of a system of concentric circles are represented by the points of a line through the common centre and perpendicular to the Oxy-plane.

4. Coaxal systems

The coaxal system defined by the two circles $\mathscr{C}$ and $\mathscr{C}'$, where $\mathscr{C}'$ is given by the equation

$$C' \equiv x^2 + y^2 - 2\xi'x - 2\eta'y + \zeta' = 0,$$

is given by the equation

$$x^2 + y^2 - 2\xi x - 2\eta y + \zeta + \lambda[x^2 + y^2 - 2\xi'x - 2\eta'y + \zeta'] = 0,$$

which, for our purpose, must be written in the form

$$x^2 + y^2 - 2\left[\frac{\xi + \lambda\xi'}{1+\lambda}\right]x - 2\left[\frac{\eta + \lambda\eta'}{1+\lambda}\right]y + \left[\frac{\zeta + \lambda\zeta'}{1+\lambda}\right] = 0.$$

The representative point is therefore

$$\left(\frac{\xi + \lambda\xi'}{1+\lambda}, \; \frac{\eta + \lambda\eta'}{1+\lambda}, \; \frac{\zeta + \lambda\zeta'}{1+\lambda}\right).$$

This point is on the line PP', where P represents $\mathscr{C}$, and P' represents $\mathscr{C}'$. Hence *the circles of a coaxal system are represented by the points of a line in E_3.* Any two distinct points of a given line in E_3 determine the line. Hence we see once more that any two distinct circles of a coaxal system determine the coaxal system.

The line PP' will meet the paraboloid Ω in two, one or no real points. Intersections with Ω correspond to zero circles, or limiting points, in the coaxal system. Hence the line representing a non-intersecting system of coaxal circles meets Ω in real

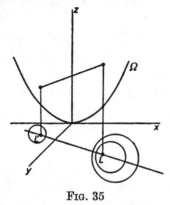

Fig. 35

points, whereas the line representing an intersecting system of coaxal circles does not meet Ω.

A pair of points in the Oxy-plane determines a coaxal system by the condition that the circles of the system should pass through both points. This coaxal system determines a line in E_3. Two lines in E_3 which meet represent coaxal systems which have one circle in common. If both systems are of intersecting types, and the common points of one are P, P', and of the other Q, Q', it follows in this case that P, P', Q, Q' are concyclic.

Bearing these facts in mind, we see that known theorems in E_3 yield theorems for coaxal systems of circles. We give two examples.

5. Deductions from the representation

Let l, m be two skew lines in E_3, and V a point which does not lie on either line. Through V we can draw a unique line n which

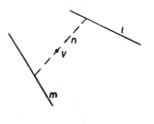

FIG. 36

will meet both l and m, that is, a unique *transversal* of l and m. This line n must lie in the plane obtained by joining l to V, and it must also lie in the plane obtained by joining m to V. These two planes intersect in a line through V, and this is the required line n. If there were two such lines n, l and m would be coplanar, contrary to hypothesis.

We translate this theorem into its plane equivalent:

Let l and m both represent intersecting coaxal systems, one defined by the points P, P', the other by the points Q, Q'. Since l and m do not meet, the coaxal systems do not have a common circle, and the four points P, P', Q, Q', are not concyclic. The point V represents a circle $\mathscr{C}$ which is not a member of either coaxal system. The line n gives a coaxal system which contains $\mathscr{C}$. If this third system is also of the intersecting type, let R, R' be the points which define it. These points lie on $\mathscr{C}$.

Since n meets both l and m, there is a circle common to the systems represented by l and n, and there is also a circle common to the systems represented by m and n. Hence P, P', R, R' are concyclic, and so are Q, Q', R, R'. The plane theorem becomes:

P, P', Q, Q' are four points which are not concyclic, and do not lie on a given circle $\mathscr{C}$. Then there is a unique pair of points R, R' on $\mathscr{C}$ such that P, P', R, R' and Q, Q', R, R' are respectively concyclic.

This theorem may, of course, be proved directly. If we draw circles through P, P', their common chords with $\mathscr{C}$ form a pencil of lines. This hint should enable the reader to complete the proof.

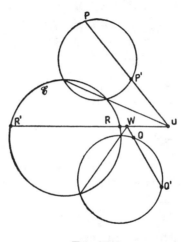

FIG. 37

Our second example is derived from a theorem in E_3 which it would take us too long to prove here:

Four lines in E_3 have two, one, or an infinity of transversals.

The proof involves the general theory of quadric surfaces, and is beyond our scope in this book. From the preceding discussion we rapidly arrive at the following theorem:

Given four pairs of points P, P', Q, Q', R, R', S, S' in a plane, the problem of finding another pair X, X' such that $PP'XX'$, $QQ'XX'$, $RR'XX'$ and $SS'XX'$ are respectively concyclic admits of two, one or an infinity of solutions.

It would not be so easy to give a direct proof of this theorem.

6. Conjugacy relations

Let $\mathscr{C}_1$ and $\mathscr{C}_2$ be two circles. Then, with the usual notation,

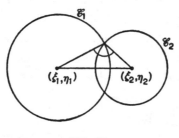

FIG. 38

we find the condition that they intersect at an angle α. The respective centres are (ξ_1, η_1) and (ξ_2, η_2), and the radii from the centres to a point of intersection enclose an angle α or $\pi - \alpha$. The cosine formula gives the relation

$$(\xi_1 - \xi_2)^2 + (\eta_1 - \eta_2)^2 = (\xi_1^2 + \eta_1^2 - \zeta_1) + (\xi_2^2 + \eta_2^2 - \zeta_2)$$
$$\pm 2(\xi_1^2 + \eta_1^2 - \zeta_1)^{\frac{1}{2}} (\xi_2^2 + \eta_2^2 - \zeta_2)^{\frac{1}{2}} \cos \alpha.$$

This becomes

$$2\xi_1\xi_2 + 2\eta_1\eta_2 - \zeta_1 - \zeta_2$$
$$= \mp 2(\xi_1^2 + \eta_1^2 - \zeta_1)^{\frac{1}{2}} (\xi_2^2 + \eta_2^2 - \zeta_2)^{\frac{1}{2}} \cos \alpha,$$

which is more useful in the form

$$(2\xi_1\xi_2 + 2\eta_1\eta_2 - \zeta_1 - \zeta_2)^2$$
$$= 4(\xi_1^2 + \eta_1^2 - \zeta_1)(\xi_2^2 + \eta_2^2 - \zeta_2)\cos^2 \alpha. \tag{1}$$

We first examine two special cases:

(i) $\alpha = 0$. The circles $\mathscr{C}_1$ and $\mathscr{C}_2$ touch. If we take the variable point $\left(\dfrac{\xi_1 + \lambda\xi_2}{1 + \lambda}, \ \dfrac{\eta_1 + \lambda\eta_2}{1 + \lambda}, \ \dfrac{\zeta_1 + \lambda\zeta_2}{1 + \lambda} \right)$ on the line

$P_1 P_2$, this point lies on Ω if and only if

$$(\xi_1 + \lambda\xi_2)^2 + (\eta_1 + \lambda\eta_2)^2 - (1 + \lambda)(\zeta_1 + \lambda\zeta_2) = 0.$$

As a quadratic in λ this becomes

$$\lambda^2(\xi_2^2 + \eta_2^2 - \zeta_2) + \lambda(2\xi_1\xi_2 + 2\eta_1\eta_2 - \zeta_1 - \zeta_2)$$
$$+ (\xi_1^2 + \eta_1^2 - \zeta_1) = 0. \tag{2}$$

The condition (1) with $\cos \alpha = 1$ is just the condition that the quadratic (2) should have equal roots. Hence, if the circles $\mathscr{C}_1$ and $\mathscr{C}_2$ touch, the line $P_1 P_2$ touches Ω. But the intersections of this line with Ω give the limiting points of the coaxal system determined by $\mathscr{C}_1$ and $\mathscr{C}_2$. We deduce that a necessary and sufficient condition that two circles touch is that the limiting points of the coaxal system they determine should coincide.

(ii) $\alpha = \frac{1}{2}\pi$. In this case (1) becomes simply

$$2\xi_1\xi_2 + 2\eta_1\eta_2 - \zeta_1 - \zeta_2 = 0. \tag{3}$$

From (2) we see that this condition implies that the roots of the quadratic are of the form $k, -k$. Hence the intersections of P_1P_2 with Ω divide P_1P_2 internally and externally in the same ratio. That is, the four points form a harmonic range, and P_1, P_2 are *conjugate* points with respect to Ω.

Hence *orthogonality implies conjugacy, and conversely.*

The locus of points conjugate, with respect to Ω, to the fixed point (ξ',η',ζ') is given by the equation

$$2x\xi' + 2y\eta' - s - \zeta' = 0. \tag{4}$$

This represents a plane in E_3, the *polar* plane of (ξ',η',ζ'). The points of this plane represent the circles orthogonal to the circle

$$x^2 + y^2 - 2\xi'x - 2\eta'y + \zeta' = 0. \tag{5}$$

The plane given by (4) meets Ω in a conic. If (x,y,s) is a point on this conic, $s = x^2 + y^2$, and since (4) is satisfied, we have

$$2x\xi' + 2y\eta' - (x^2 + y^2) - \zeta' = 0,$$

which is the same equation as (5). The projection of the point (x,y,s) on to the plane Oxy is the point $(x,y,0)$, and we have proved

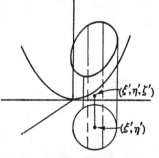

FIG. 39

that *the projection of the conic in which the polar plane of a point in E_3 meets Ω is precisely the circle represented by the point.*

We may see this from another point of view. The points of the conic cut on Ω represent *point* circles, that is, circles of zero radius, orthogonal to the circle given by (5). It is easily verified that if a point circle is orthogonal to a given circle, the centre of the point circle must lie *on* the given circle. Hence the projection of the conic on to the plane Oxy is bound to give the circle itself.

The polar plane of the point
$$\left(\frac{\xi_1 + \lambda\xi_2}{1+\lambda}, \ \frac{\eta_1 + \lambda\eta_2}{1+\lambda}, \ \frac{\zeta_1 + \lambda\zeta_2}{1+\lambda}\right),$$
which, as λ varies, describes the line P_1P_2, is
$$2\xi\left(\frac{\xi_1 + \lambda\xi_2}{1+\lambda}\right) + 2\eta\left(\frac{\eta_1 + \lambda\eta_2}{1+\lambda}\right) - \zeta - \left(\frac{\zeta_1 + \lambda\zeta_2}{1+\lambda}\right) = 0.$$
This may be written in the form
$$2\xi\xi_1 + 2\eta\eta_1 - \zeta - \zeta_1 + \lambda(2\xi\xi_2 + 2\eta\eta_2 - \zeta - \zeta_2) = 0,$$
which shows that the polar plane always passes through a fixed line, given by the intersection of the polar planes of P_1 and of P_2. If we call this line l', and call P_1P_2 l, any point on l is conjugate to every point of l', and therefore any point on l' is conjugate to every point of l. Two such lines are called *polar* lines, and they represent conjugate coaxal systems. If l meets Ω in the points P, P', the polar planes of these points are the respective tangent planes at the points to Ω, and these tangent planes will contain l'. If l meets Ω, and so represents a non-intersecting system of coaxal circles, l' will not meet Ω.

The tangent plane at P represents the circles orthogonal to the point-circle represented by P, that is, the circles which *pass through* the projection of P on to the Oxy-plane. Hence, l' represents the circles which pass through the limiting points of the coaxal system represented by l.

Again, the projection of l on to the plane Oxy gives the line of centres of the coaxal system represented by l. Hence *polar lines project into perpendicular lines*.

We can easily find a tetrahedron in E_3 which is self-polar: that is, every face is the polar plane of the opposite vertex. Choose any point P not on Ω, and let π be its polar plane. Choose any point Q in π not on Ω, and let π' be its polar plane. π' contains P and meets π in a line l. Choose any point R on l not on Ω. Its polar plane contains P and Q and meets l in S. Then it is clear that $PQRS$ is a self-polar tetrahedron.

Since polar lines project into perpendicular lines, and the opposite edges of $PQRS$ are polar lines, we see that *if four circles are mutually orthogonal, each being orthogonal to the other three, then their centres form a triangle and its orthocentre*. It can be shown that one vertex of a self-polar tetrahedron must lie *inside* Ω. Hence, one of the four mutually orthogonal circles must be imaginary.

7. Circles cutting at a given angle

We now return to the general relation (1) of the previous section, and see that the points representing circles which cut a fixed circle $\mathscr{C}'$ at an angle α or $\pi - \alpha$ satisfy the equation

$$4(x^2 + y^2 - \varepsilon)\,(\xi'^2 + \eta'^2 - \zeta')\cos^2\alpha$$
$$= (2x\xi' + 2y\eta' - \varepsilon - \zeta')^2. \qquad (1)$$

This represents a quadric surface, and the form of its equation shows that it has *ring-contact* with Ω along the intersection of Ω with the plane

$$2x\xi' + 2y\eta' - \varepsilon - \zeta' = 0. \qquad (2)$$

If the reader is unfamiliar with the idea of ring-contact he need not be disturbed, since it is only the algebraic form of (1) which will concern us. We know already that the curve in which the plane given by (2) meets Ω projects into the circle $\mathscr{C}'$ itself.

We wish to investigate the circles which cut three given circles $\mathscr{C}_1$, $\mathscr{C}_2$ and $\mathscr{C}_3$ at angles α or $\pi - \alpha$. To do this, we must examine the common intersections of the three quadric surfaces:

$$4(x^2 + y^2 - \varepsilon)\,(\xi_r^2 + \eta_r^2 - \zeta_r)\cos^2\alpha$$
$$= (2x\xi_r + 2y\eta_r - \varepsilon - \zeta_r)^2 \qquad (r = 1, 2, 3).$$

For the sake of brevity we write these equations as

$$4(x^2 + y^2 - \varepsilon)k_r\cos^2\alpha = X_r^2 \qquad (r = 1, 2, 3), \qquad (3)$$

where k_r is a constant, and $X_r = 0$ is the polar plane of (ξ_r, η_r, ζ_r). If we equate the common value of $x^2 + y^2 - \varepsilon$ in the three equations, we obtain the equations

$$\frac{X_1^2}{k_1} = \frac{X_2^2}{k_2} = \frac{X_3^2}{k_3}.$$

We note that these equations do not depend on $\cos\alpha$. If we take them in pairs, the system is equivalent to

$$\begin{cases} \sqrt{k_2}\,X_1 \pm \sqrt{k_1}\,X_2 = 0, \\ \sqrt{k_3}\,X_2 \pm \sqrt{k_2}\,X_3 = 0. \end{cases} \qquad (4)$$

Taking all the possible combinations of signs, we have four sets of two linear equations in x, y, ε. Since two planes determine a line, the solutions of (4) lie on four straight lines. Since $X_1 = X_2 = X_3 = 0$ automatically satisfies (4), these four lines all pass through the point determined by the intersection of the polar planes of (ξ_r, η_r, ζ_r) $(r = 1, 2, 3)$. This point represents the circle orthogonal to the three given circles.

Each one of the lines determined by (4) meets each one of the three quadrics given by (3) in the same two points. Hence there are *eight* solutions of the equations, for a given value of cos α, and we have proved the following theorem:

There are eight circles which intersect three given circles at a given angle α or π — α. As the angle α varies the circles vary in four coaxal systems with a common circle, the circle orthogonal to the three given circles.

The case α = 0 gives the solution of the problem of Apollonius (see chapter I, §10). Since each one of the lines determined by (4) gives two solutions of (3), each one of the four coaxal systems contains two of the circles which satisfy the given conditions.

8. Representation of inversion

Let A, A' be inverse points in a circle $\mathscr{C}$. Then $\mathscr{C}$ is orthogonal to any circle through A and A', and $\mathscr{C}$ is a member of the coaxal system which is determined by the limiting points A, A'. Let P represent the circle $\mathscr{C}$. The zero circles centres A, A' are represented by points on Ω. By the remark above, P is collinear with these two points on Ω.

Since P is regarded as fixed, and A, A' as variable, we have the following method for finding the inverse, in a given circle $\mathscr{C}$, of a given curve, $\mathscr{D}$, say:

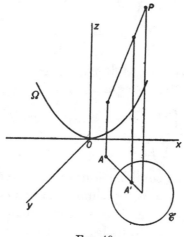

FIG. 40

Erect perpendiculars to the plane Oxy at the points of $\mathscr{D}$, and let $\mathscr{E}$ be the curve in which these perpendiculars meet Ω. The cone of lines joining P to the points of $\mathscr{E}$ meets Ω in a further curve $\mathscr{E}'$. Project $\mathscr{E}'$ down on to the plane Oxy. The resulting curve $\mathscr{D}'$ is the inverse of $\mathscr{D}$ in $\mathscr{C}$.

We remark that this construction shows at once that the eight circles which were discussed in the preceding section are inverse in pairs in the common orthogonal circle of the three circles. Since each of the three given circles inverts into itself with respect to the common orthogonal circle, this result is to be expected.

If the curve $\mathscr{D}$ to be inverted is a circle, we know that the curve $\mathscr{E}$ obtained by projecting it on to Ω is a plane section. The cone of lines joining P to the points of $\mathscr{E}$ is a quadric cone, and since the inverse of a circle is another circle, this cone meets Ω again in a further plane section $\mathscr{E}'$. Again, a circle, the circle of inversion and the inverse circle are coaxal, so that we obtain the theorem:

A quadric cone which meets Ω in one plane section will meet it again in another plane section, and the poles of the two plane sections are collinear with the vertex of the cone.

The more advanced reader will realize that this is a *projective* theorem, and is therefore true for any quadric Ω.

9. The envelope of a system

We may measure the complexity of a system of circles by the complexity of the curve in E_3 which represents the system. From this point of view a coaxal system of circles is the simplest kind of system, since a line is the simplest curve in E_3. A coaxal system of circles has no *envelope*: that is, there is no curve touched by all circles of the system. For intersecting systems we may, if we wish, say that the envelope consists of the two points common to all circles of the system. But interesting envelopes only appear when we consider *conic systems of circles*.

A conic system of circles is one represented by a conic in E_3. A conic is a plane curve, and this plane will have a pole with respect to Ω. This pole represents a circle which is orthogonal to all the circles of the conic system. Again, the projection of the conic on to the plane Oxy is, except in one special case, a conic, and gives the locus of centres of circles of the conic system.

Hence the centres of circles of a conic system lie on a conic, and all circles of the system are orthogonal to a fixed circle.

These two properties serve to define a conic system, since the second property restricts the representative points to lie in a plane, and the intersection of a conical cylinder and a plane is a conic.

We now find the envelope of a conic system of circles. Let the curve touched by all circles of the system be $\mathscr{E}$, and suppose that the circle $\mathscr{C}$ touches it at the point A. A circle of the system

Fig. 41

"near" to $\mathscr{C}$ will intersect $\mathscr{C}$ in a point which is "near" to A. In the limit the intersection of $\mathscr{C}$ and a neighbouring circle of the conic system includes the point of contact of $\mathscr{C}$ with $\mathscr{E}$. It can be shown that all circles of the system touch the locus of ultimate intersections of neighbouring circles, so that all circles of the system touch the envelope at *two* points.

Let C be the conic in E_3 which represents the conic system of circles. If P is the point on C which represents $\mathscr{C}$, the tangent to C at P represents the coaxal system defined by $\mathscr{C}$ and its neighbouring circle. The common points of this coaxal system are the points on $\mathscr{E}$ we wish to determine. Now the line in E_3 polar to the tangent to C at P (with respect to Ω) represents the *conjugate* coaxal system, whose limiting points are the points on $\mathscr{E}$ required. If we find the intersections of this polar line with Ω, and project down on to the plane Oxy, we have the points we want.

Therefore, *to obtain the envelope of the system of circles represented by a curve C in E_3, we find the curve in which lines polar to the tangents of C (with respect to Ω) meet Ω, and project this curve orthogonally on to the (x,y)-plane.*

The method applies to any curve C, but we are interested in the case in which C is a conic. Then the lines polar to the tangents of C all pass through a fixed point, the pole of the plane of C, and generate a quadric cone. Hence, the envelope is the orthogonal projection of the curve in which a quadric cone meets Ω.

Now two quadrics intersect in a curve which is cut by a plane in four points, these points being the four intersections of the respective conics in which the plane cuts the quadrics. Such a curve is called a *quartic* curve. If the quadric intersecting Ω has the equation

$$Q \equiv ax^2 + by^2 + cz^2 + 2fyz + 2gzx + 2hxy + 2ux$$
$$+ 2vy + 2wz + d = 0,$$

the orthogonal projection of the intersection with Ω is obtained by replacing z, wherever it occurs in the equation, by $x^2 + y^2$. We therefore obtain the curve

$$ax^2 + by^2 + c(x^2 + y^2)^2 + 2fy(x^2 + y^2) + 2g(x^2 + y^2)x$$
$$+ 2hxy + 2ux + 2vy + 2w(x^2 + y^2) + d = 0.$$

This curve has double points at the circular points at infinity, and is called a *bicircular quartic curve*. The envelope of a conic system of circles is therefore a bicircular quartic curve.

The quartic curve on Ω is the intersection of two quadrics $\Omega = 0$ and $Q = 0$, and therefore the pencil of quadrics $\Omega + \lambda Q = 0$ passes through it. A pencil of quadrics contains four cones. Hence four cones intersect Ω in the quartic curve whose projection gives the envelope. One of these cones has appeared already, its generators being the polar lines of the tangents to the conic C which specified the conic system of circles. Since each quadric cone can be derived as the set of polar lines of tangents to a conic, we see that the bicircular quartic curve may be regarded as the envelope of *four* systems of circles.

Again, a conic in E_3 meets Ω in four points. Hence a conic system of circles contains four point-circles. Since a point-circle

$$(x - a)^2 + (y - b)^2 = 0$$

may also be written in the form $(i = \sqrt{-1})$:

$$[(x - a) + i(y - b)][(x - a) - i(y - b)] = 0,$$

a point-circle may also be regarded as the isotropic lines through the centre (a,b). If these touch a given curve, the point (a,b) is, by definition, a focus of the curve.

Hence the bicircular quartic we have obtained has sixteen foci.

It is easy to see that any bicircular quartic may be obtained in this way, and therefore has the properties described above, but we shall not pursue the matter any further.

10. Some further applications

Let us suppose that we are given a plane algebraic curve, of equation $f(x,y) = 0$. This curve is touched by an infinity of circles through the point $(0,0)$. If such a circle is

$$x^2 + y^2 - 2\xi x - 2\eta y = 0,$$

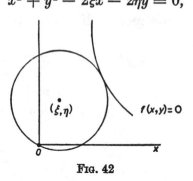

FIG. 42

a relation $g(\xi,\eta) = 0$ holds; that is, *the centres of circles through $(0,0)$ which touch a given curve $f(x,y) = 0$ lie on a curve $g(x,y) = 0$.*

We may call this latter curve *the circle-tangential equation* of the given curve. We consider some examples.

If the curve $f(x,y) = 0$ consists merely of the point (a,b), the circle through $(0,0)$ has to pass through (a,b). The locus of centres

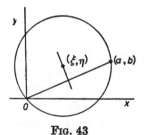

FIG. 43

of such circles is the perpendicular bisector of the join of (a,b) and $(0,0)$.

If $f(x,y) = 0$ represents a line, the circle-tangential equation is

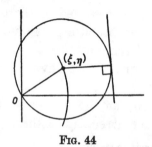

FIG. 44

the locus of a point which moves so that its distance from a fixed point is always equal to its distance from a fixed line.   Thus the locus is a parabola, with (0,0) as focus and the line as directrix.

Finally, we consider the case when $f(x,y) = 0$ is the circle
$$x^2 + y^2 - 2\xi'x - 2\eta'y + \zeta' = 0.$$

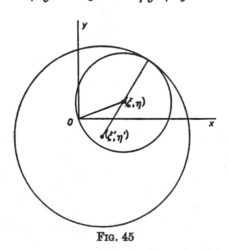

FIG. 45

There are two cases:

(i) If (0,0) lies inside the circle ($\zeta' < 0$), the circle-tangential equation is the locus of a point which moves so that the *sum* of its distances from two fixed points, (0,0) and (ξ',η'), is constant.   The locus is therefore an ellipse with these points as foci.

(ii) If (0,0) lies outside the circle ($\zeta' > 0$), it is the *difference* of the distances from (0,0) and (ξ',η') which is constant.   The circle-tangential equation is then a hyperbola with these points as foci.

We may apply the methods of the preceding section to finding the circle-tangential equation of a given curve $f(x,y) = 0$.   When applied to the above examples, we shall obtain some properties of the paraboloid Ω.

Let us suppose then that $g(x,y) = 0$ is the circle-tangential equation of $f(x,y) = 0$.   Since all the circles we are considering pass through (0,0), *the curve $g(x,y) = 0$ is itself the representative curve in E_3 of a system of circles which have $f(x,y) = 0$ as envelope.* Applying the method of the last section, we find the polar lines of the tangents to $g(x,y) = 0$.   We obtain a cone of lines, vertex (0,0).   This cone meets Ω in a curve of which $f(x,y) = 0$, the envelope, is the orthogonal projection on to the (x,y)-plane.

Since we are given $f(x,y) = 0$, to obtain $g(x,y) = 0$ we project

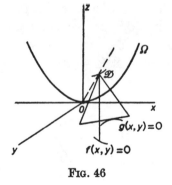

FIG. 46

the given curve up on to Ω, obtaining a curve $\mathscr{D}$, say. The tangent planes to Ω at the points of $\mathscr{D}$ meet the (x,y)-plane in the tangents of the required circle-tangential equation. This follows from the fact that the polar line of a line passing through $(0,0)$ is the intersection of the polar planes of $(0,0)$ and the point where this line meets Ω again. The polar plane of $(0,0)$ is the (x,y)-plane, and the polar plane of any other point on Ω is also the tangent plane at the point.

We apply this method to the examples given above. For the circle-tangential equation of a point (a,b), we consider the intersection with $z=0$ of the tangent plane to Ω at the point (a,b, a^2+b^2). This intersection must be the perpendicular bisector of the join of (a,b) to $(0,0)$.

We now operate with the circle $x^2+y^2-2\xi'x-2\eta'y+\zeta'=0$. When we project this up on to Ω, we obtain a plane section given by the intersection of Ω and the plane

$$2\xi'x + 2\eta'y - z - \zeta' = 0.$$

Tangent planes to Ω at the points of this section all pass through the pole of the plane, which is the point (ξ',η',ζ'), and meet $z=0$ in tangents to the curve given by the intersection with $z=0$ of the tangent cone to Ω from (ξ',η',ζ').

The tangent cone is easily derived from Eq. (2), §6, chapter II, and is given by the equation

$$4(\xi'^2+\eta'^2-\zeta')(x^2+y^2-z)-(2\xi'x+2\eta'y-\zeta'-z)^2=0.$$

If we put $z=0$ in this equation, we obtain

$$4(\xi'^2+\eta'^2-\zeta')(x^2+y^2)-(2\xi'x+2\eta'y-\zeta')^2=0.$$

Our previous investigation tells us that this is *a conic with foci at* $(0,0)$ *and* (ξ',η'), *and an ellipse or a hyperbola according as* ζ' *is negative or positive.*

The presence of a focus at (0,0) is manifest, since the equation may be written as

$$x^2 + y^2 = \frac{(\xi'^2 + \eta'^2)}{(\xi'^2 + \eta'^2 - \zeta')} \left[\frac{2\xi'x + 2\eta'y - \zeta'}{2\sqrt{(\xi'^2 + \eta'^2)}} \right]^2.$$

The corresponding directrix is the intersection with $s = 0$ of the polar plane of (ξ',η',ζ'). We have found then that the tangent cone to Ω from the point (ξ',η',ζ') meets $s = 0$ in a conic with foci at $(0,0)$ and (ξ',η'), and that this conic is an ellipse or hyperbola according as ζ' is negative or positive.

11. Some anallagmatic curves

We conclude this chapter by finding the equations of curves of a given order which are anallagmatic (self-inverse) in a given circle.

Let the circle have representative point (ξ,η,ζ). As we saw in §8, page 36, points which are inverse in the circle

$$x^2 + y^2 - 2\xi x - 2\eta y + \zeta = 0$$

are represented by points on Ω collinear with the point (ξ,η,ζ).

A plane through (ξ,η,ζ) is given by the equation

$$p(x - \xi) + q(y - \eta) + r(s - \zeta) = 0.$$

This plane intersects Ω in a conic, the orthogonal projection of which is the circle

$$p(x - \xi) + q(y - \eta) + r(x^2 + y^2 - \zeta) = 0.$$

By construction, this circle is self-inverse, and therefore orthogonal to the given circle. It is the most general circle orthogonal to the given circle.

A quadric cone with vertex at (ξ,η,ζ) has the equation

$$a(x - \xi)^2 + 2h(x - \xi)(y - \eta) + b(y - \eta)^2$$
$$+ 2g(x - \xi)(s - \zeta) + 2f(y - \eta)(s - \zeta) + c(s - \zeta)^2 = 0,$$

and therefore the quartic curves anallagmatic in the given circle are obtained by merely substituting $s = x^2 + y^2$ in this equation. They are bicircular quartics.

To obtain the anallagmatic *cubic* curves, the quartic curve of intersection of the quadric cone vertex (ξ,η,ζ) with Ω must have one point at infinity on the axis, $x = y = 0$, of Ω. Therefore, one generator of the cone must be parallel to $x = y = 0$. We therefore put $c = 0$ in the above equation to obtain anallagmatic cubics.

This process can easily be extended.

CHAPTER III

OUR aim in this chapter is to discuss the Poincaré model of hyperbolic non-Euclidean geometry. This sounds more formidable than it really is. Circles play a leading part in the discussion, as we shall see, and so do the elementary properties of complex numbers. We begin by recapitulating these properties.

1. Complex numbers

Complex numbers are expressions of the form $a + bi$, where a and b are any real numbers, the rules of calculation for these complex numbers being the same as for real numbers, plus the rule that $i.i = i^2 = -1$.

Hence, we have
$$(a + bi) + (c + di) = (a + c) + (bi + di) = (a + c) + (b + d)i,$$
and
$$\begin{aligned}(a + bi)(c + di) &= ac + bci + adi + bdi^2 \\ &= ac + bci + adi - bd \\ &= (ac - bd) + (bc + ad)i.\end{aligned}$$

We say that a is the *real part*, bi the *imaginary part* of the complex number $a + bi$. A complex number is only zero if its real part and imaginary part are simultaneously zero. Hence, $a + bi = c + di$ if and only if $a = c$ and $b = d$.

2. The Argand diagram

If, in a plane, a system of rectangular Cartesian axes be taken, the complex number $z = x + iy$ may be represented by the point with coordinates (x,y). In this way we set up a correspondence between the complex numbers and points of the plane. In particular, the numbers $x + i.0$ correspond to points of Ox, which is therefore called the *real axis*. Such numbers are indistinguishable from ordinary real numbers x.

This method of representing complex numbers is called the Argand diagram. The plane thus put into correspondence with the set of complex numbers is called *the plane of the complex variable,* and we shall speak of *the complex number z* or of *the point*

s, this point s being precisely the point which corresponds to the number s.

It is sometimes useful to think of the point s as a *vector* $\overrightarrow{Os}$ with components (x,y). We recall that two vectors are equal if they have the same components, that is to say if the arrows which represent them are parallel, in the same sense, and of equal length, without necessarily starting from the same origin.

We see at once that these vectors obey the parallelogram law

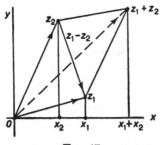

FIG. 47

of addition, since if $s_1 = x_1 + iy_1$, and $s_2 = x_2 + iy_2$,
$$s = s_1 + s_2 = x_1 + x_2 + i(y_1 + y_2).$$
The interpretation of subtraction is equally simple:

If $s = s_1 - s_2$, and therefore $s_1 = s + s_2$, s must be represented by the vector $\overrightarrow{s_2 s_1}$.

3. Modulus and argument

If the length of the vector $\overrightarrow{Os}$ be denoted by ρ, and the angle

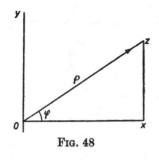

FIG. 48

through which Ox must be rotated to lie along $\overrightarrow{Os}$ by φ, we have
$$x = \rho \cos \varphi, \qquad y = \rho \sin \varphi, \qquad \rho = \sqrt{(x^2 + y^2)}.$$
ρ is called the *modulus* of the complex number, and φ, which is determined to a multiple of 2π, the *argument*.

It follows that
$$z = x + iy = \rho(\cos \varphi + i \sin \varphi).$$
If now z_1 and z_2 are two complex numbers,
$$z_1 = \rho_1(\cos \varphi_1 + i \sin \varphi_1),$$
$$z_2 = \rho_2(\cos \varphi_2 + i \sin \varphi_2),$$
their product $z_1 z_2 = \rho_1 \rho_2 (\cos (\varphi_1 + \varphi_2) + i \sin (\varphi_1 + \varphi_2))$, as is easily seen by direct multiplication. Hence, to multiply two complex numbers, you multiply their moduli and add their arguments.

For division, let $z = z_1/z_2$, $(z_2 \neq 0)$, so that
$$z_1 = z z_2.$$
Using the first rule, it follows that the quotient of two complex numbers has, for modulus, the quotient of their moduli, and, for argument, the difference of their arguments.

4. Circles as level curves

Let a, b, z be any three complex numbers. We denote by (zab) the ratio

$$\frac{z - a}{z - b} = (zab).$$

The modulus of this complex number is the quotient of the moduli of $z - a$ and $z - b$. It is therefore the quotient

$$az^\dagger/bz \,.$$

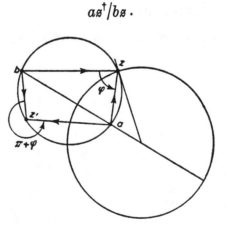

FIG. 49

---

† az denotes the distance between the points a and z in the Argand diagram.

As for its argument, by the above it is the angle through which the vector $z - b$ must be turned to give it the direction of the vector $z - a$.   It is therefore the angle between the vectors $\overrightarrow{az}$ and $\overrightarrow{bz}$, oriented as in Fig. 49.

If we regard a and b as fixed, and z as variable, we may enquire into the nature of the curves along which mod (zab) and arg (zab) are respectively constant.   These are the *level curves* of the respective functions.   Both turn out to be circles.

If we have $az/bz = $ constant, the ratio of the powers of the point z with respect to the point circles a and b is constant.   Therefore z moves on a circle of the coaxal system determined by the point circles a,b.   This coaxal system has these point circles as limiting points.

If the angle between $\overrightarrow{az}$ and $\overrightarrow{bz}$ is constant, z moves on an arc of a certain circle passing through a and b.   The other arc, the arc complementary to the first, corresponds to an argument $\varphi + \pi$. If φ takes all possible values, we obtain all circles of the coaxal system through a and b, including the radical axis ab, when $\varphi = \pi$ (this gives points between a and b) and when $\varphi = 0$ (this gives points outside ab).

For all values of the modulus and argument of (zab) we therefore obtain all circles of two conjugate coaxal systems as level curves.

5. The cross-ratio of four complex numbers

Let a, b, c, d be four complex numbers.   By definition, the cross-ratio of the four numbers in the given order is the **ratio** $(acd)/(bcd)$.   We write

$$\frac{(acd)}{(bcd)} = (ab,cd),$$

so that

$$(ab,cd) = \frac{(a - c)/(a - d)}{(b - c)/(b - d)}.$$

This is formally the same definition as that which occurs in projective geometry.

The *modulus* of the cross-ratio is equal to the quotient of the moduli of (acd) and (bcd).   Hence, for this modulus to be equal to

1, it is necessary and sufficient for (acd) and (bcd) to have the same modulus. By what we have seen above, a and b must therefore lie on the same circle of the coaxal system which has c,d as limiting points.

The *argument* of the cross-ratio is equal to the difference of the arguments of (acd) and (bcd). The cross-ratio therefore has zero argument if (acd) and (bcd) have the same argument, which is the case if a and b are on the same arc of a circle bounded by c and d.

The argument of the cross-ratio is equal to π if the arguments

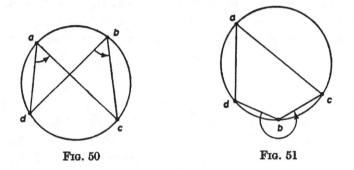

FIG. 50   FIG. 51

of (acd) and (bcd) differ by π, which is the case if a and b are on complementary arcs of a circle through c and d.

Now, a number is real if its argument is 0 or π. Hence we have the theorem:

For the cross-ratio (ab,cd) to be real, it is necessary and sufficient that the four points a, b, c, d should lie on a circle (or straight line).

More generally, we find the locus of z such that the cross-ratio (za,bc) has (i) a constant modulus; and (ii) a constant argument.

Since $(za,bc) = (zbc)/(abc)$, the modulus and argument of (za,bc) will be constant if the respective modulus and argument of (zbc) are constant.

(i) The locus of points z such that mod (zbc) is constant is a circle of the coaxal system with limiting points b and c.

(ii) The locus of points z such that arg (zbc) is constant is an arc of a circle limited by b and c.

We deduce that if a, b, c are three distinct complex numbers, u any given complex number, then *there is a unique number z such that*

$$(za,bc) = u.$$

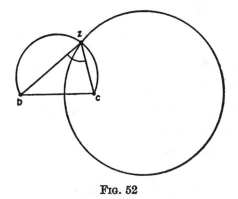

FIG. 52

For u has a definite modulus and argument, and z is given as the unique intersection of a circle and an arc of a circle, as shown in Fig. 52.

This important property is worth verifying algebraically. If $(za,bc) = u$, then we must have $(zbc) = (abc)u$. Write $(abc)u = k$. Then $(z - b)/(z - c) = k$, which gives

$$z = \frac{b - kc}{1 - k}.$$

We conclude this section by giving an interesting geometrical

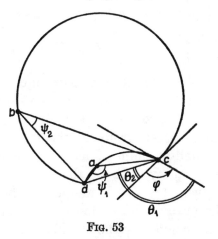

FIG. 53

interpretation to the cross-ratio (ab,cd) of any four complex numbers. If ψ_1 is the argument of (acd), and ψ_2 that of (bcd), we know that the argument φ of (ab,cd) satisfies the equation

$$\varphi = \psi_1 - \psi_2.$$

Draw the circles through a, c, d and b, c, d respectively, and let Θ_1, Θ_2 be the angles made by cd with the tangents at c to these circles. Then by the alternate segment theorem, $\psi_1 = \Theta_1$ and $\psi_2 = \Theta_2$, so that $\varphi = \Theta_1 - \Theta_2$. Hence φ is the angle between the two circles, and we have the theorem:

The argument of the cross-ratio (ab,cd) is equal to the angle between the two circles passing through a, c, d and b, c, d respectively.

6. Möbius transformations of the z-plane

These are transformations of the plane of the complex variable which map any point z on a point w according to the formula

$$w = \frac{Az + B}{Cz + D},\qquad(1)$$

where A, B, C, D are four complex constants, and $AD - BC \neq 0$.

We exclude the case $A/C = B/D$, because then $w = A/C$ for all values of z, and all points z transform into a single point w.

The reader is probably already familiar with transformations of type (1) in projective geometry, where they are called *bilinear* or *projective* transformations, and transform points of a line. The same algebra proves the theorem:

A Möbius transformation leaves cross-ratios invariant.

In other words, if a, b, c, d transform respectively into a', b', c', d', then $(ab,cd) = (a'b',c'd')$.

We make some important deductions from this theorem.

I. If $w = \dfrac{Az + B}{Cz + D}$, $(AD - BC \neq 0)$ is a Möbius transformation, and (a,a'), (b,b'), (c,c') are three pairs of corresponding points, then the equation of the transformation may be written in the form

$$(wa',b'c') = (za,bc).\qquad(2)$$

This follows immediately from the theorem. If z is any point, and w the corresponding point under the Möbius transformation, then the pair (z,w) satisfy (2). On the other hand we know that, z having been given, there is a unique w which can make the cross-ratio $(wa',b'c') = (za,bc)$. Hence the equation (2) does, in fact, map z on the w given by the Möbius transformation.

II. Conversely, let a, b, c be three distinct complex numbers, and a', b', c' a second set of three distinct complex numbers. Then the equation

$$(wa',b'c') = (za,bc)$$

is the equation of a Möbius transformation which maps a on a', b on b' and c on c'.

This follows immediately by writing out both sides of the equation, and substituting $z = a, b, c$ in turn. We find that $w = a'$, b', c' in turn.

III. A Möbius transformation transforms circles into circles. (It is understood that straight lines come into the category of circles.)

We shall give two proofs of this important theorem.

In the first place let z be any point of the circle determined by the three points a, b, c. Then the cross-ratio

$$(za,bc) = \text{real number.}$$

Let z' be the transform of z. Then

$$(z'a',b'c') = (za,bc) = \text{real number,}$$

and therefore z' lies on a circle through a', b', c'.

The second proof will be given in the next section. We conclude this section with:

IV. A Möbius transformation conserves the angle of intersection of two circles.

Let b and c be the points common to two circles $\mathscr{C}_1$ and $\mathscr{C}_2$, a_1 a third point on $\mathscr{C}_1$ and a_2 a third point on $\mathscr{C}_2$. We have seen (see page 50) that the angle between the two circles is given by the argument of the cross-ratio (a_1a_2,bc). But this argument is conserved, since the cross-ratio itself is conserved.

A second proof of this theorem will also be given later.

7. A Möbius transformation dissected

The reader may ask, at this stage, whether there is any connection between Möbius transformations and inversion. We show that there is.

We may write

$$w = \frac{Az + B}{Cz + D} = \frac{A}{C} + \frac{BC - AD}{C(Cz + D)},$$

by simple division, or

$$w = \frac{A}{C} + \frac{BC - AD}{C^2} \cdot \frac{1}{z + D/C}.$$

We now carry out the transformation on z in several stages:

(i) Let $$z_1 = z + \frac{D}{C};$$

(ii) Let
$$z_2 = \frac{1}{z_1} \; ;$$

(iii) Let
$$z_3 = \frac{BC - AD}{C^2} \cdot z_2 \; ;$$

(iv) Then
$$w = \frac{A}{C} + z_3.$$

Each stage can be interpreted geometrically, and this we proceed to do.

Stage (i) is equivalent to a *translation* of z in a definite direction and through a definite distance, given by the vector D/C.

Stage (ii) is equivalent to inversion in the unit circle, followed by reflection in the real axis. For with an obvious notation,

$$r_2(\cos \Theta_2 + i \sin \Theta_2) = \frac{1}{r_1(\cos \Theta_1 + i \sin \Theta_1)}$$

$$= \frac{1}{r_1} (\cos (- \Theta_1) + i \sin (- \Theta_1)),$$

and therefore $r_2 = 1/r_1$, so that $r_1 r_2 = 1$, and
$$\Theta_2 = - \Theta_1.$$

Stage (iii), *multiplication* by a fixed complex number is equivalent to a dilatation by a fixed amount from the origin, followed by a definite rotation about the origin. For if $(BC - AD)/C^2 = \rho(\cos \Theta + i \sin \Theta)$,

$$r_3 = \rho r_2,$$
and
$$\Theta_3 = \Theta_2 + \Theta.$$

Stage (iv) is another translation.

We see that after each one of these sub-transformations circles are still circles (or straight lines), and that the angle of intersection of two circles is unchanged. Hence we have additional proofs of two of the theorems of the preceding section.

Another important deduction we can make from our dissection of a Möbius transformation is the following:

If a circle $\mathscr{C}$ is transformed into a circle $\mathscr{C}'$, the interior of $\mathscr{C}$ is mapped on the interior or the exterior of $\mathscr{C}'$.

We mean, of course, that every point inside $\mathscr{C}$ is mapped on points inside $\mathscr{C}'$, or every point inside $\mathscr{C}$ is mapped on points outside $\mathscr{C}'$.

The only transformation which can turn a circle inside-out is (ii), since this is an inversion followed by a reflection. The inverse of a circle with respect to a centre of inversion inside the circle turns the circle inside out. The other transformations which come into the dissection preserve insides and outsides.

If the circle $\mathscr{C}'$ is a straight line, the inside of $\mathscr{C}$ is mapped on one of the half-planes bounded by the line, the outside of $\mathscr{C}$ on the other.

8. The group property

The Möbius transformations of the z-plane form a group. Before we prove this, we give the definition of a *group of transformations*.

A set of transformations is said to form a group if it has the following properties:

(a) Every transformation T of the set possesses an inverse transformation which is also in the set. If this inverse transformation be denoted by T', the transformation effected by carrying out first T and then T' is the identical transformation, which leaves everything unchanged.

(b) If two transformations T_1, T_2 both belong to the set, the transformation obtained by carrying out first T_1 and then T_2 also belongs to the set.

That the Möbius transformations of the z plane form a group is a theorem easily proved by algebra. Alternatively we can proceed as follows:

The transformation inverse to
$$(za,bc) = (z'a',b'c'),$$
that is, the one which maps z' on z, is given by the equation
$$(z'a',b'c') = (za,bc).$$
Now let M_1 be a Möbius transformation which maps a on a', b on b', c on c'. It is determined by these three pairs of corresponding points. Let M_2 be a Möbius transformation which maps a' on a'', b' on b'' and c' on c''. It is also determined by these three pairs of corresponding points.

M_1 is given by the equation $(z'a',b'c') = (za,bc)$;

M_2 is given by the equation $(z''a'',b''c'') = (z'a',b'c')$.

The result of M_1 followed by M_2 is the transformation
$$(z''a'',b''c'') = (za,bc),$$
which is a Möbius transformation.

If a set of transformations, all of which belong to a group G of transformations, themselves form a group, we call this a *subgroup* of the group G. An important example of a group of transformations is afforded by the group of displacements in a Euclidean plane.

We imagine a fixed plane, and another like a large sheet of paper covering it. The paper is moved, with any possible figures on it, from one position over the plane to another. We consider only the initial and final positions of a motion or displacement. The path does not concern us. Suppose A, B, points of the

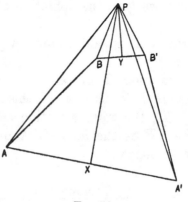

Fig. 54

plane in its first position, become A', B' in its final position. Join AA', BB', and let their perpendicular bisectors PX, PY meet in P. Evidently the triangles PAB, $PA'B'$ are congruent. Hence the displacement envisaged can be effected by a rotation round P.

If the right bisectors were parallel, then AA', BB' would be parallel, and the displacement would be a *translation*, every point of the plane moving through the same distance and in the same direction.

If we follow a given displacement by another, the result is a displacement, since only initial and final positions are considered. The inverse of a given displacement is merely that displacement which brings the moving plane back to its original position. Hence the displacements, or Euclidean motions of a plane, form a group.

The set of translations of the plane evidently form a subgroup of the group of displacements.

We may also regard the displacement of one plane over another as a *mapping* of the points of a Euclidean plane into itself which is one-to-one, transforms lines into lines, preserves angles, and also orientation. It is this notion which is the basis of *applicability* in Euclidean geometry. In the next section we investigate a group of transformations which, in our non-Euclidean geometry, corresponds to the group of Euclidean motions.

9. Special transformations

We are interested in a certain subgroup of the group of Möbius transformations, namely that subgroup which consists of all Möbius transformations which map the inside of a given circle on itself. Let us be more explicit.

We choose the circle with centre at the origin and radius $= 1$. We call the circumference ω, and the circular area bounded by ω we call Ω. The points of ω are not included in Ω.

There are Möbius transformations which transform Ω into itself. We shall call these M-transformations. For example:

the identical transformation $w = z$;
the transformation $w = (\cos \varphi + i \sin \varphi)z$;

this gives a rotation of the plane, through an angle φ, about the origin.

If an M-transformation maps Ω on itself, the dissection of §7 shows that the boundary ω is also mapped on itself. But the converse, as we know from the simple example $w = 1/z$, is not true.

It is easy to see that M-transformations form a group, which is a subgroup of the group of all Möbius transformations. For, if a transformation maps Ω on itself, this is also true for the inverse transformation; and if two transformations M_1, M_2 both map Ω on itself, this is also true for the transformation obtained by first carrying out M_1 and then M_2.

We shall now restrict ourselves to these M-transformations and to Ω, the interior of the unit circle, centre at the origin. We shall show that Ω *can be regarded as a non-Euclidean space, in which the M-transformations are the group of non-Euclidean motions*.

10. The fundamental theorem

We recall one of the characteristic properties of the group of displacements of a plane (over itself). Let A be any point in the plane, α a direction through it, X any other point in the plane and

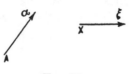

ξ a direction through it. Then there is a unique displacement of the plane which maps A on X and α on ξ.

We show that the M-transformations have the same property in Ω. The theorem to be proved is:

Let X and A be any two points of Ω, ξ and α directions through X and A respectively. Then there is a unique M-transformation which maps (X,ξ) on $(A,α)$.

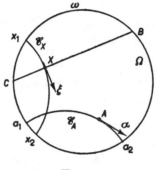

FIG. 56

We note in the first place that, through any point X, and touching any given line through the point, there passes exactly one circle orthogonal to ω. This is immediate by inversion, or may be proved as follows: draw the perpendicular at X to the given line, and let B, C be the points in which it cuts ω. Among the circles of the pencil through B and C, two are evidently orthogonal to the circle sought for: ω itself and the line BC. Hence the circle required is of the coaxal system conjugate to that defined by ω and the line BC. It therefore has B and C as limiting points, and is uniquely determined, since it passes through X.

If we care to regard it as an Apollonius circle, it is the locus of a point which moves so that the ratio of its distances from two fixed points B and C is always equal to XB/XC.

Call this circle $\mathscr{C}_X$, and let x_1 and x_2 be its intersections with ω. In the same way, there passes through A a unique circle $\mathscr{C}_A$ orthogonal to ω and tangent at A to the direction α. Let a_1, a_2 be its intersections with ω.

Now, since M-transformations map ω on itself, they necessarily map circles orthogonal to ω on to circles orthogonal to ω, being Möbius transformations.

Hence any M-transformation which maps (X,ξ) on (A,α) must necessarily map $\mathscr{C}_X$ on $\mathscr{C}_A$, since $\mathscr{C}_X$ will be mapped on a circle through A, with direction α, and orthogonal to ω, and there is only one such circle, namely $\mathscr{C}_A$.

Let us now suppose that the points of intersection of $\mathscr{C}_X$ and $\mathscr{C}_A$ with ω are so numbered that the direction ξ is that of x_1 towards x_2 in Ω, and the direction α is that of a_1 towards a_2 in Ω. The M-transformation we are looking for must map x_1 on a_1, x_2 on a_2 and X on A. These three pairs of corresponding points determine a unique Möbius transformation. Our proof will therefore be complete if we show it is an M-transformation, mapping Ω into itself. This is clear, since the unique Möbius transformation maps the circle through x_1 and x_2 orthogonal to $\mathscr{C}_X$, that is ω, on the circle through a_1 and a_2 orthogonal to $\mathscr{C}_A$, which is again ω. It therefore maps ω on ω. On the other hand it maps Ω on Ω, since it maps the interior point X of Ω on the point A of Ω. The unique Möbius transformation is therefore an M-transformation.

Corollary: Let a and b be two points of Ω, $\mathscr{C}_a$ a circle orthogonal to ω passing through a, $\mathscr{C}_b$ a circle orthogonal to ω passing through b. There-exist precisely *two* M-transformations mapping a on b and $\mathscr{C}_a$ on $\mathscr{C}_b$. To a given orientation on $\mathscr{C}_a$ the one transforma-

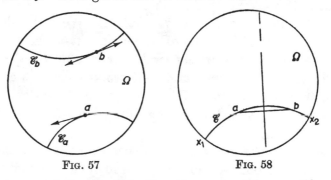

FIG. 57   FIG. 58

tion maps one of the possible directions of motion on $\mathscr{C}_b$, and the second transformation maps the other.

In particular, let a, b be two points of Ω. There exists a unique M-transformation which permutes a and b (that is, which maps a on b and b on a).

To prove this, we note in the first instance that there exists a unique circle orthogonal to ω and passing through a and b. Call this circle $\mathscr{C}$. An M-transformation interchanging a and b must transform $\mathscr{C}$ into itself.

By the above, there are only two M-transformations which map a on b and $\mathscr{C}$ on itself. The one which conserves orientation on $\mathscr{C}$ cannot map b on a, since then the sense of a towards b in Ω would be transformed into the sense of b towards a in Ω.

Let M' therefore be the M-transformation which maps a on b and alters the orientation of $\mathscr{C}$. We show that M' maps b on a.

Let x_1 and x_2 be the intersections of $\mathscr{C}$ and ω, and a' the transform of b by M'. We want to prove that $a' = a$. Since M' changes the orientation of $\mathscr{C}$, x_1 is mapped on x_2, x_2 on x_1. Since a is also mapped on b, and b on a', the cross-ratio property gives

$$(x_1 x_2, ab) = (x_2 x_1, ba').$$

But, for any cross-ratio,

$$(x_1 x_2, ab) = (x_2 x_1, ba).$$

Hence

$$(x_2 x_1, ba) = (x_2 x_1, ba'),$$

and therefore

$$a' = a.$$

11. The Poincaré model

We can now discuss Poincaré's model of a hyperbolic non-Euclidean geometry. The term " non-Euclidean " refers to the fact that in this geometry Euclid's parallel axiom does not obtain.

Our space is *the interior Ω of the circle ω*. We shall call this the *h*-plane. A point of Ω is a point, or *h*-point, in our new geometry. But *our h-lines are the arcs of circles (or straight lines) inside Ω, limited by ω and orthogonal to ω*. These are the arcs considered in our previous theorems. If a point in Ω lies on one of these arcs, we shall say that the corresponding *h*-point lies on the corresponding *h*-line.

We see that the ordinary incidence theorems of Euclidean geometry hold in our *h*-geometry. Through two distinct points of Ω there passes a unique circle orthogonal to ω. Hence, *through two distinct h-points there passes a unique h-line*.

Two circles orthogonal to ω intersect in at most one point which is inside Ω. Hence, two h-lines intersect in at most one h-point. We shall consider the pairs of h-lines which do not intersect later.

Since arcs of circles inside Ω orthogonal to ω constantly recur, we call them ω-arcs.

The angle between two ω-arcs will be, by definition, the h-angle between the corresponding h-lines.

The relation of equality between the angles of ω-arcs will imply the relation of equality, or h-equality, between h-angles.

We now come to the more difficult notion of *h-equality* between h-segments. An *h-segment* is naturally the portion of an ω-arc bounded by two points of Ω.

In elementary geometry two segments are equal if one can be applied to the other. This involves a displacement, or motion, of the plane. Such a motion transforms lines into lines, and conserves angles. In our h-geometry the equivalent to this group of motions is the group of M-transformations, which transform h-lines into h-lines, and conserve angles. We therefore call an M-transformation of Ω an *h-displacement of the h-plane*, and say that two h-segments AB and CD are h-equal, written

$$AB \overset{h}{=} CD,$$

if there is an h-motion which transforms AB into CD. Since M-transformations form a group, the relation of h-equality satisfies the usual axioms of equivalence:

(i) $AB \overset{h}{=} AB$:

(ii) if $AB \overset{h}{=} CD$, then $CD \overset{h}{=} AB$:

(iii) if $AB \overset{h}{=} CD$, and $CD \overset{h}{=} EF$, then

$$AB \overset{h}{=} EF.$$

Again, since there is an M-transformation mapping the segment ab of an ω-arc on the segment ba, we also have

$$AB \overset{h}{=} BA.$$

The ordinary theorems on congruence in Euclidean geometry have an equivalent in our h-geometry. For example, let us prove the theorem:

In two h-triangles ABC and $A'B'C'$, the equalities $AB \overset{h}{=} A'B'$, $AC \overset{h}{=} A'C'$ and $\angle CAB \overset{h}{=} \angle C'A'B'$ imply the equality $\angle ABC \overset{h}{=} \angle A'B'C'$.

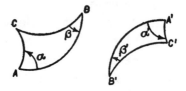

FIG. 59

To prove this we must show that if two triangles ABC, $A'B'C'$ (made up of ω-arcs) are such that $\alpha = \alpha'$ (Fig. 59), and there exists an M-transformation which maps AB on $A'B'$, and an M-transformation which maps AC on $A'C'$, then $\beta = \beta'$.

Suppose at first that the triangles ABC, $A'B'C'$ have the same orientation. The proof will consist in showing that there is an M-transformation which maps the ω-triangle ABC on the ω-triangle $A'B'C'$.

Since $AB \overset{h}{=} A'B'$, there exists a (unique) M-transformation which maps A on A' and B on B'. Since this transformation preserves angles in magnitude and orientation, it maps the ω-arc AC on the ω-arc $A'C'$ in the direction $A'C'$. This M-transformation is the only one which maps A on A' and the ω-arc AC in the direction AC on the ω-arc $A'C'$ in the direction $A'C'$. Hence, it is the one which maps A on A' and C on C', since we know such a transformation exists.

We have therefore found an M-transformation which maps the ω-triangle ABC on the ω-triangle $A'B'C'$. This transformation conserves angles, and therefore $\beta = \beta'$.

If the orientations of the ω-triangles ABC, $A'B'C'$ are different, we take *the geometrical image* of $A'B'C'$ in any diameter of Ω. Call the transformed triangle $A''B''C''$. Then the ω-triangles ABC, $A''B''C''$ have the same orientation, and the proof applies. Since the ω-triangle $A''B''C''$ arises from $A'B'C'$ by a reflection, the theorem follows.

12. The parallel axiom

So far we have shown that there is a complete equivalence between our *h*-geometry and that part of Euclidean geometry which is developed before the introduction of the parallel axiom. We now come to the parting of ways !

We could say that two ω-arcs which do not intersect shall be called *h-parallel h-lines*. Through a point *A* outside any ω-arc we can draw an infinity of ω-arcs which do not cut the given ω-arc. This would lead to there being an infinity of *h*-lines *h*-parallel

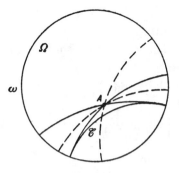

Fig. 60

to a given *h*-line through a given *h*-point outside the *h*-line.

We notice that there are two ω-arcs through *A* which *touch* the given ω-arc, the respective points of contact being on ω. These two ω-arcs separate the ω-arcs through *A* into two classes: those which cut the given ω-arc, and those which do not.

We prefer to reserve the term *h-parallel* for the two ω-arcs which touch the given ω-arc at a point of ω, and therefore now have, instead of Euclid's parallel axiom, the theorem:

Through any h-point A taken outside an h-line there pass precisely two h-parallels to the given h-line.

Hence, the term *non-Euclidean* geometry. The existence of a model for non-Euclidean geometry shows that the parallel axiom is *independent* of the other axioms, and *cannot be deduced from them*. Much fruitless effort has gone into attempts to *prove* the parallel axiom, and even today there exist organizations of people devoted to proving the axiom. Needless to say, these devotees are not shaken in their faith by mathematical arguments.

13. Non-Euclidean distance

We know how to compare two *h*-segments for equality. We

now want to obtain a *distance-function* $D[ab]$ for any two h-points a, b which satisfies the conditions:

$$D[ba] = D[ab] \geqslant 0;$$
$$D[ab] = 0 \qquad \text{if and only if } b = a;$$
$$D[ab] + D[bc] \geqslant D[ac],$$

with equality if and only if b lies on the h-line ac between a and c.

We also want this distance function to be invariant under h-motions of the h-plane.

Let the h-line ab cut ω in the points α, β. Then

$$D[ab] = \left| \log (\alpha\beta, ab) \right|$$

is a suitable distance-function. It evidently satisfies the first two conditions, and also the condition that it be invariant under h-motions of the h-plane.

For $\qquad D[ba] = \left| \log (\alpha\beta, ba) \right|$

$$= \left| \log \frac{1}{(\alpha\beta, ab)} \right| = \left| -\log (\alpha\beta, ab) \right|$$

$$= D[ab],$$

and, by definition, $D[ab] \geqslant 0$. It can only $= 0$ if $(\alpha\beta, ab) = 1$, when we must have $b = a$. Again, if any M-transformation of Ω maps the points α, β, a, b on α', β', a', b' respectively, we know that α', β' will be on ω, and that α', β', a', b' all lie on an h-line. Since also $(\alpha\beta, ab) = (\alpha'\beta', a'b')$,

$$D[ab] = D[a'b'].$$

If a, b, c are three points, in this order, on an h-line which cuts ω in α, β, we have, identically,

$$(\alpha\beta, ac) = (\alpha\beta, ab) . (\alpha\beta, bc).$$

As the points a, b, c are ordered as shown, the three logarithms $\log (\alpha\beta, ac)$, $\log (\alpha\beta, ab)$ and $\log (\alpha\beta, bc)$ all have the same sign. Therefore

$$D[ac] = D[ab] + D[bc].$$

To prove the full triangle inequality is more difficult, but we remark that since the parallel axiom is not used in Euclid's earlier theorems, we can apply the first twenty-eight propositions of Euclid's first book, without change, to our h-plane. We then obtain theorems concerning the congruence of triangles, as we have seen, the theorem that the greatest side of a triangle is opposite the greatest angle, and lastly, the triangle inequality:

$$D[ab] + D[bc] > D[ac],$$

if a, b, c are not, in this order, on an h-line.

It is possible to go much further into the study of non-Euclidean geometry, but we do not propose to do so in this book. The reader may find that his interest in the axioms of Euclidean geometry has been aroused, and he cannot do better than to study these afresh, and then see what are the essential differences between the non-Euclidean geometry we have been discussing and Euclidean geometry.

CHAPTER IV

THIS chapter discusses the solution of a classical problem: to show that, of all closed plane curves of a given perimeter, the circle encloses the greatest area. This property of a circle is usually referred to as its *isoperimetric* property.

1. Steiner's enlarging process

Steiner gave a simple geometric construction by which, given any closed convex plane curve K which is not a circle, it is possible to devise a new curve K^* which is also plane and closed, has the *same* perimeter, but contains a *greater* area than K.

It follows that K cannot be the solution of the isoperimetric problem.

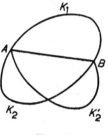

FIG. 61

The method is as follows. Choose on K two points A and B which bisect the perimeter; that is, the two arcs into which A and B divide K have equal length. If one arc be K_1, and if this, together with AB, bounds a curve of area F_1, while the other arc K_2 and AB bound a curve of area F_2, then we assume that $F_1 \geqslant F_2$. The area of the curve is $F = F_1 + F_2$.

We now wipe out the arc K_2, and substitute in its place the arc K_2' which is obtained from K_1 by reflection in the line AB. The closed curve bounded by K_1 and K_2' evidently has the same perimeter as K. Call this new curve K'. Its area is $F' = 2F_1$, and we have

$$F \leqslant F'.$$

Since the equality sign may hold in this relation, we have not yet completed the process. By hypothesis, K was not a circle. Hence, we may certainly choose A and B so that neither of the arcs K_1 and K_2 is a semicircle. Therefore K' is not a circle.

Hence we can find a point C distinct from A and B on K' so that $\angle ACB = \gamma$ is not a right angle. Let D be the image of C in AB. If we cut out from the area bounded by K' the quadrangle $ACBD$, four *lunes* remain, as indicated below.

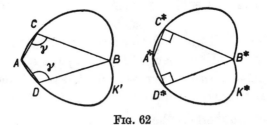

FIG. 62

We imagine that the sides of the quadrangle are bars, hinged at A, B, C and D, and that the lunes are fastened to these bars. We now move the bars about the hinges to a position where the angles at C and D are both right angles. Call the new quadrangle $A^*C^*B^*D^*$. Then the symmetric curve K^*, which circumscribes the new position of the quadrangle, is the one we are seeking.

In fact, the perimeter of K^* is equal to that of K', being made up of four arcs each congruent to the corresponding arc of K'. The difference in the areas enclosed by K' and K^* is equal to the difference in the areas of $ACBD$ and $A^*C^*B^*D^*$, since the lunes are unchanged in area. If F^* be the area bounded by K^*,

$$F^* - F' = CA.CB\,(1 - \sin\gamma) > 0.$$

Hence $F^* > F'$, and therefore

$$F^* > F,$$

so that our new curve K^* has the *same* perimeter, but encloses a *greater* area.

2. Existence of a solution

Does the process we have just described *prove* the isoperimetric property of the circle ? We have shown that if K is a plane closed curve, not a circle, then the enlarging process always produces another closed plane curve K^* of the *same* perimeter but enclosing a *greater* area. Hence, the curve K cannot be a solution of the isoperimetric problem.

If, among all plane closed curves of a given perimeter, we can find one whose area $\geq$ the area of every other one, this curve must be a circle.

However, this does not solve the problem, since we must prove that such a curve exists. The Steiner construction indicates a method for proceeding *towards* a solution, but this is not enough, and we must digress for a moment.

If the perimeter of a given plane closed curve be L, we shall see later on that the area enclosed, F, satisfies the inequality

$$F < L^2.$$

Hence, the set of all plane closed curves with given perimeter L gives rise to a set of numbers F *bounded from above*. Such a set of numbers has a *least* upper bound Λ, say. Then Λ has the following characteristic properties:

(i) $\Lambda \geq$ all numbers in the set;

(ii) for any $\epsilon > 0$, there exists at least one number in the set $> \Lambda - \epsilon$.

If the least upper bound Λ is a *member* of the set of numbers F, the isoperimetric property of the circle is proved. But once again, this would have to be proved. The set of numbers $1/2$, $2/3, \ldots, n/(n + 1), \ldots$ has 1 as a least upper bound, but 1 is not a member of the set, so that in a bounded infinite set of numbers there does not necessarily exist a greatest one.

We are therefore up against a serious difficulty, and, although it can be done, we shall not prove the isoperimetric property of the circle by a continued use of Steiner's enlarging process.

3. Method of solution

So far we have not discussed what we mean *precisely* by the perimeter of a closed curve, and the area enclosed by it. This remains to be done. What we shall do in this section is to enumerate the points in our proof of the isoperimetric property, giving the details later.

Let Λ be the perimeter and Φ the area of an equilateral polygon with an even number of vertices. Then we can show that

$$\Lambda^2 - 4\pi\Phi > 0.$$

It follows from this that if K is any closed, rectifiable, continuous curve of perimeter L and area F, then

$$L^2 - 4\pi F \geq 0.$$

To show this, we inscribe in K an equilateral polygon V^* with an even number of vertices, and choose the sides ρ of V^* so small that the perimeter Λ^* and area Φ^* of V^* differ respectively from the perimeter L and area F of K by as small a quantity as we please. If we had

$$L^2 - 4\pi F < 0,$$

we could construct V^* so that

$$\Lambda^{*2} - 4\pi\Phi^* < 0,$$

and so arrive at a contradiction. Hence

$$L^2 - 4\pi F \geqslant 0.$$

If K is a circle, $L = 2\pi r$, $F = \pi r^2$, and

$$L^2 - 4\pi F = 0.$$

If K is any other continuous, closed and rectifiable curve, we can deduce from it, by the Steiner procedure, a curve K' of perimeter L' and area F' such that

$$L = L' \quad \text{and} \quad F < F'.$$

Since

$$L'^2 - 4\pi F' \geqslant 0,$$

it follows that we must have

$$L^2 - 4\pi F > 0.$$

Hence we have the following result:

Let K be a continuous, closed and rectifiable plane curve of perimeter L and area F. Then

$$L^2 - 4\pi F \geqslant 0,$$

and equality holds if and only if K is a circle, described positively.

This proves the isoperimetric property of the circle:

Amongst all plane curves of a given perimeter, the circle, described positively, has the greatest area.

The remainder of this chapter is devoted to the discussion of the concepts *perimeter* and *area* and to the proof of the statements made above.

4. Area of a polygon

Before we can discuss the concept *area enclosed by a curve*, we must be clear about the area of a polygon.

We take the usual system of rectangular Cartesian coordinates,

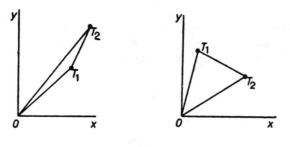

FIG. 63

and suppose that T_1, T_2 are two points with coordinates (x_1,y_1) and (x_2,y_2).   Then the area of the triangle OT_1T_2 is given by the formula:

$$\text{area } OT_1T_2 = \tfrac{1}{2}(x_1y_2 - x_2y_1).$$

If we have $n+1$ points: $T_i(x_i,y_i)$ $(i = 1, \ldots, n+1)$, then, by the area of the *polygon* $OT_1T_2 \cdots T_{n+1}$ we understand the sum:

$$\text{area } OT_1T_2 + \text{area } OT_2T_3 + \cdots + \text{area } OT_nT_{n+1}$$

$$= \tfrac{1}{2} \sum_{k=1}^{n} (x_ky_{k+1} - x_{k+1}y_k).$$

We assume that T_{n+1} and T_1 coincide.   The area is now

$$\tfrac{1}{2} \sum_{k=1}^{n} (x_ky_{k+1} - x_{k+1}y_k) \qquad (x_{n+1} = x_1;\ y_{n+1} = y_1).$$

A simple calculation shows that this is *independent of the position of the origin*, and we call it *the area of the closed polygon* $T_1T_2 \cdots T_n$.

By a *polygon*, we mean a finite number of points T_1, T_2, $\cdots$, T_n which are not necessarily distinct, but are ordered.   The area is also invariant under a rotation of the axes, and we can therefore say that *two similarly oriented congruent polygons have the same area*.   If we reverse the sense in which the polygon is described, we change the sign of the area.

If two polygons have a sequence of vertices in common which are described in opposite senses for the two polygons, their areas can be added, and we obtain the area of the resulting polygon. For example:

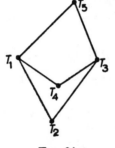

Fig. 64

area $(T_1T_2T_3T_4)$ + area $(T_1T_4T_3T_5)$ = area $(T_1T_2T_3T_5)$.
The *perimeter* of a polygon is a simpler concept. We have

$$\text{perimeter}\,(T_1T_2\cdots T_n) = \sum_1^n T_kT_{k+1},$$

where the length of each side is taken positively.

5. Regular polygons

We assume that we are given an equilateral polygon V with an even number of vertices n, and perimeter Λ. If n and Λ are fixed, when is the area of the polygon a maximum?

An equilateral polygon is said to be *regular* if its vertices lie on a circle and if, on traversing the circumference of the circle once, all the vertices are traversed just once.

We can show that the answer to the question posed above is: *the area of the polygon is a maximum when it is regular, and traversed positively.*

We prove this in two stages. By applying the Steiner enlarging process, we complete the first stage, and can show that *no polygon which is not regular can have a maximum area.*

The second stage is an existence proof: *Amongst all equilateral polygons with a given number n of vertices and a given perimeter Λ, there is one whose area $\geqslant$ area of any other.*

We shall leave the proof of the first stage to the reader, as it is quite straightforward. To prove the second stage, we first show that:

The areas of all the polygons we are considering are bounded from above.

We may assume that T_1 is at the origin. Then every length $OT_k \leqslant \Lambda/2$, for O and T_k are connected by two paths, made up of sides of the polygon, whose total length is Λ. One of these paths must therefore certainly be of length $\leqslant \Lambda/2$. Therefore the length of the straight segment $OT_k \leqslant \Lambda/2$, by the triangle inequality.

Since also $T_k T_{k+1} = \Lambda/n$, we have

$$|\text{area of triangle } OT_k T_{k+1}| < \Lambda^2/4n.$$

On the other hand:

$$|\Phi| = |\text{area } (T_1 T_2 \cdots T_n)| \leqslant n \max |\text{area } (OT_k T_{k+1})|,$$

and therefore we have the desired result:

$$|\Phi| < \tfrac{1}{4}\Lambda^2.$$

The bounded set of the numbers denoting the areas Φ of all the polygons we are considering therefore has a least upper bound Φ_0, and we wish to show that Φ_0 is in the set; that is, that there exists at least one polygon with Φ_0 for the measure of its area.

To prove this, we invoke a classical theorem of function theory. If we consider the expression for the area

$$\Phi = \tfrac{1}{2} \sum_{k=1}^{n} (x_k y_{k+1} - x_{k+1} y_k),$$

it is evidently a continuous function of the $2n$ variables $x_1, y_1, \ldots, x_n, y_n$. If we keep T_1 at the origin, we can easily designate a closed area of the plane within which the polygon always lies. The square bounded by $x = -\Lambda, x = \Lambda, y = -\Lambda, y = \Lambda$ will do. Φ is a continuous function of $x_1, y_1, \ldots, x_n, y_n$ within and on the boundary of the product space given by:

$$x_1 = \pm \Lambda, y_1 = \pm \Lambda, \ldots, x_n = \pm \Lambda, y_n = \pm \Lambda.$$

But we also have the equations

$$(x_k - x_{k+1})^2 + (y_k - y_{k+1})^2 = \Lambda^2/n^2 \qquad (k = 1, \ldots, n).$$

In the product space these equations define a bounded, closed set of points. It is a classical theorem that Φ attains its maximum on this set.

Hence if Φ is the area of a *non-regular* equilateral polygon with an even number of vertices, and Φ_0 is the area of the *regular* polygon, described positively, with the same perimeter and number of sides, then

$$\Phi < \Phi_0.$$

If r is the radius of the circle circumscribing the regular polygon, then

$$\Lambda_0 = 2nr \sin \frac{\pi}{n},$$

$$\Phi_0 = n \left(\tfrac{1}{2} r^2 \sin \frac{2\pi}{n} \right) = n r^2 \sin \frac{\pi}{n} \cos \frac{\pi}{n},$$

so that

$$\Lambda_0^2 - \left(4n \tan \frac{\pi}{n} \right) \Phi_0 = 0.$$

For any equilateral polygon with n vertices and perimeter $\Lambda = \Lambda_0$, we have, by the above proof,

$$\Lambda^2 - \left(4n \tan \frac{\pi}{n} \right) \Phi > 0,$$

so that, in all cases,

$$\Lambda^2 - \left(4n \tan \frac{\pi}{n} \right) \Phi \geqslant 0.$$

We replace this by a weaker inequality which does not involve the number of sides.

Write $\dfrac{\pi}{n} = p$, then

$$4n \tan \frac{\pi}{n} = 4\pi \tan p / p.$$

But since $0 < p < \dfrac{\pi}{2}$, we know that $\tan p > p$, so that

$$4\pi \tan p / p > 4\pi,$$

and our inequality becomes

$$\Lambda^2 - 4\pi \Phi > 0.$$

This is the fundamental inequality for an equilateral polygon with an even number of sides used in §3.

6. Rectifiable curves

We now define the *perimeter* of a closed curve.

If $a \leqslant t \leqslant b$, the continuous functions $x(t)$, $y(t)$ give a parametric representation

$$x = x(t), \, y = y(t)$$

of a *continuous curve* K with initial point A $(t = a)$ and end point B $(t = b)$. We suppose that there is no subinterval $\alpha \leqslant t \leqslant \beta$ in which both functions x, y are constant.

We now take on K the points $T_1, T_2, \ldots, T_{n-1}$ which correspond to the parametric values

$$a < t_1 < t_2 < \cdots < t_{n-1} < b.$$

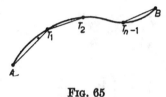

FIG. 65

We join the points $A, T_1, T_2, \ldots, T_{n-1}, B$ by straight lines in order, and obtain the sum

$$\Lambda = AT_1 + T_1T_2 + \cdots + T_{n-1}B,$$

giving the length of the inscribed path $AT_1T_2 \ldots T_{n-1}B$.

If this sum is always bounded, we say that K is *rectifiable*, and the least upper bound of the set of positive numbers Λ is called the *arc length* L of K.

From the property of the least upper bound, we always have

$$\Lambda \leqslant L,$$

and for any $\epsilon > 0$ there always exist values Λ satisfying

$$\Lambda > L - \epsilon.$$

An immediate deduction from the definition of arc length is the following: if two points A and B are joined by a continuous rectifiable curve K which is distinct from the line AB, then the arc length of K exceeds the length AB.

For if K contains a point M which is not on the line AB, then the arc length L of K satisfies the inequality

$$L \geqslant AM + MB > AB.$$

If M is a point on K corresponding to a value m, where $a < m < b$, it can be proved, in an evident notation, that

$$L_a^b = L_a^m + L_m^b.$$

We leave this proof to the reader.

We now consider a *closed* continuous curve K:

$$x = x(t), \qquad y = y(t), \qquad (a \leqslant t \leqslant b),$$
$$x(a) = x(b), \qquad y(a) = y(b).$$

We can remove the restriction that $a \leqslant t \leqslant b$ if we decide that $x(t), y(t)$ are to have the *period* $b - a$; that is,

$$x(t + b - a) = x(t),$$
$$y(t + b - a) = y(t),$$

for all values of t. By a linear substitution,

$$\varphi = pt + q,$$

we can transform the interval $a \leqslant t \leqslant b$ into the interval $0 \leqslant \varphi \leqslant 2\pi$. Replacing t by φ, we have two continuous functions:

$$x = x(\varphi), \qquad y = y(\varphi),$$

of period 2π. If, besides this, we put

$$\xi = \cos \varphi, \qquad \eta = \sin \varphi,$$

(ξ, η) describes the unit circle. Every point of this circle corresponds, to a multiple of 2π, to a unique value of φ, and this value of φ corresponds to a unique point on K. Hence we may say:

By a continuous closed curve we mean the continuous mapping of the points of a circle. Such a mapping is not necessarily one-to-one.

We now define the *perimeter* of a closed curve. We merely use the previous definition of arc length when $A = B$:

$$L = L_a^b.$$

L is independent of the choice of the point $A = B$; that is,

$$L_a^b = L_{a+c}^{b+c}.$$

By the rule for the addition of arcs, to prove this we must show that

$$L_a^{a+c} + L_{a+c}^b = L_{a+c}^b + L_b^{b+c},$$

which reduces to

$$L_a^{a+c} = L_b^{b+c}.$$

Since $a = b$, this is evident.

We can sum up for a closed curve K and say:

Inscribe in K a polygon whose vertices follow each other in the correct cyclic order. If the least upper bound L of the perimeters $\varLambda$ of all these inscribed polygons is finite, we say that K is rectifiable, and L is its perimeter.

7. Approximation by polygons

We return once again to a rectifiable arc K:

$$x = x(t), \qquad y = y(t), \qquad (a \leqslant t \leqslant b),$$

whose initial point A and end point B do not necessarily coincide. We show that the arc length L of K can be approached by the lengths $\varLambda$ of inscribed straight line paths in a *uniform* manner. Let V be such a path with vertices corresponding to the parametric values

$$a = t_0 < t_1 < t_2 < \cdots < t_n = b.$$

We have the following theorem:

Given any $\epsilon > 0$ we can always determine $\delta > 0$ so that the length $\varLambda$ of an inscribed path V differs from the length L of K by less than ϵ, that is

$$L - \varLambda < \epsilon,$$

as soon as all the parametric differences

$$t_k - t_{k-1} < \delta \qquad (k = 1, \ldots, n).$$

From the definition of L we can find a path V' inscribed in K whose length Λ' satisfies the inequality

$$L - \Lambda' < \epsilon/2.$$

Let $A = T_0, T_1', \ldots, T_m' = B$ be the vertices of V'. If we choose δ so that

$$\delta < t_k' - t_{k-1}' \qquad (k = 1, \ldots, m),$$

and inscribe in K a path V for which

$$t_k - t_{k-1} < \delta \qquad (k = 1, \ldots, n),$$

then, between two consecutive vertices T_{k-1}', T_k' of V' there is at least one vertex T_r of V. The path from A to B which contains the vertices of V and of V' has a length Λ'' such that

$$\Lambda'' \geqslant \Lambda', \text{ and therefore } L - \Lambda'' < \epsilon/2.$$

Suppose now that T_k' lies between T_r and T_{r+1}. Since the continuity of $x(t)$, $y(t)$ implies uniform continuity, for a suitably small value of δ both the distances $T_r T_k'$ and $T_k' T_{r+1}$ will be smaller than an arbitrarily given $\eta > 0$. Then

$$\Lambda'' - \Lambda = \Sigma(T_r T_k' + T_k' T_{r+1} - T_r T_{r+1})$$
$$< \Sigma(T_r T_k' + T_k' T_{r+1}) < 2\eta m.$$

Hence

$$L - \Lambda < \frac{\epsilon}{2} + 2\eta m.$$

We can choose δ so that

$$\eta < \frac{\epsilon}{4m},$$

and we then have

$$L - \Lambda < \epsilon,$$

which is our goal.

We can make the successive segments of our paths equal, if we wish. Describe a circle with A, the initial point of K, as centre, and radius ρ so small that K does not lie entirely within this circle. The first point of K, that is the point which has the smallest t-value and lies on the circumference of the circle, we call T_1. With T_1 as centre and radius ρ we describe a circle, and call the first point following T_1 which lies on its circumference T_2. We

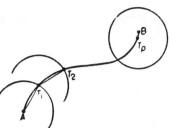

FIG. 66

continue in this way, and obtain a path consisting of equal segments of length ρ inscribed in K.

As the length of this path cannot exceed L, after a finite number of steps we reach a point T_p such that the arc of K between T_p and B lies within the circle radius ρ and centre T_p. If we now join up the points $A, T_1, T_2, \ldots, T_p, B$ in order, we obtain a path V^* inscribed in K whose first p sides are of length ρ, whilst the last side $\leqslant \rho$. We now prove:

ρ *can be taken so small that all the parametric differences* $t_{k+1} - t_k$ $(k = 0, 1, \ldots, p)$ *corresponding to the vertices* $A = T_0, T_1, \ldots,$ $T_p, T_{p+1} = B$ *are smaller than an arbitrary positive number* δ:

$$t_{k+1} - t_k < \delta \qquad (k = 0, 1, \ldots, p).$$

To see this, we first construct a path V' with vertices $A = T_0', T_1', \ldots, T_{m+1}' = B$ for which $t_{k+1}' - t_k' < \delta/2$. Since, by hypothesis, no subinterval $\alpha \leqslant t \leqslant \beta$ corresponds to a single point of K, we can choose V' so that no two consecutive points T_k' and T_{k+1}' coincide. If we now take 2ρ smaller than the smallest segment of V', there must lie between two successive vertices T' at least one vertex T, and therefore all differences $t_{k+1} - t_k < \delta$, as desired.

Since $T_p B \leqslant \rho$, the path V^* is not, in general, equilateral. If we replace this last segment by a path (not inscribed) of two or three sides each of length ρ, we obtain, instead of V^*, an equilateral path $'V^*$ with an *even* number of sides. This path is not inscribed, in the previous sense.

But if we take $A = B$, so that K is closed, the difference between the areas of the polygons V^* and $'V^*$ is equal to the area of the triangle (or quadrangle) constructed on $T_p B$ as base, and is therefore $< 4\rho^2$, by a previous result (see page 70), whilst the perimeters differ by less than 3ρ. Hence we can say:

Given $\epsilon > 0$ we can determine ρ, the side of an equilateral polygon V^* with an even number of vertices approximating to K so that the relation

$$L - \Lambda^* < \epsilon$$

holds between the length of K and the perimeter of V^*.

8. Area enclosed by a curve

Once again we assume K to be closed, continuous and rectifiable:

$$x = x(t), \qquad y = y(t) \qquad (a \leqslant t \leqslant b),$$
$$x(a) = x(b), \qquad y(a) = y(b).$$

We inscribe in K a polygon $A = T_0, T_1, \ldots, T_n = B = A$, corresponding to parametric values

$$a = t_0 < t_1 < \cdots < t_n = b.$$

The area Φ of this polygon is given by the formula:

$$2\Phi = \sum_{k=1}^{n} \left[x(t_{k-1})y(t_k) - y(t_{k-1})x(t_k) \right].$$

We wish to show that, given any $\epsilon > 0$ we can find $\delta > 0$ such that if

$$t_k - t_{k-1} < \delta \qquad (k = 1, \ldots, n),$$

then
$$F - \Phi < \epsilon,$$

where F is a number which we shall call *the area enclosed by K*.

We cannot do this without a further study of the functions $x(t), y(t)$. We know that K is rectifiable. Hence

$$\sum_{k=1}^{n} \left[\{x(t_k) - x(t_{k-1})\}^2 + \{y(t_k) - y(t_{k-1})\}^2 \right]^{\frac{1}{2}}$$

is bounded, for all possible polygons inscribed in K. Since

$$[\{x(t_k) - x(t_{k-1})\}^2 + \{y(t_k) - y(t_{k-1})\}^2]^{\frac{1}{2}}$$
$$\geqslant \left| x(t_k) - x(t_{k-1}) \right|,$$
$$\geqslant \left| y(t_k) - y(t_{k-1}) \right|,$$

we see that the function $x(t)$ has the property that the sum

$$\sum_{k=1}^{n} \left| x(t_k) - x(t_{k-1}) \right|$$

is always bounded. Similarly for $y(t)$. Such functions are called *functions of bounded variation*. Hence, if K is rectifiable, the continuous functions $x(t), y(t)$ must both be of bounded variation. Since

$$[\{x(t_k) - x(t_{k-1})\}^2 + \{y(t_k) - y(t_{k-1})\}^2]^{\frac{1}{2}}$$
$$\leqslant \left| x(t_k) - x(t_{k-1}) \right| + \left| y(t_k) - y(t_{k-1}) \right|,$$

the condition is also sufficient.

To facilitate the handling of the expression for Φ, we prove the following theorem:

Every continuous function $f(t)$ of bounded variation can be written as the difference

$$f(t) = \varphi(t) - \psi(t)$$

of two continuous and monotone increasing functions.

Let $\alpha \leqslant t \leqslant \beta$ be any subinterval of $a \leqslant t \leqslant b$. We divide it up as shown:

$$\alpha = t_0 < t_1 < \cdots < t_n = \beta,$$

and form the sum

$$\sum_{k=1}^{n} \left| f(t_k) - f(t_{k-1}) \right|.$$

By hypothesis, this sum is bounded. Let the least upper bound for all subdivisions of $\alpha \leqslant t \leqslant \beta$ be denoted by

$$S_\alpha^\beta f.$$

Actually, this is the arc-length of the continuous curve

$$x = f(t), \qquad y = 0$$

which lies on the x-axis, between $t = \alpha$ and $t = \beta$.
Now

$$\varphi(t) = S_\alpha^t$$

is a monotone increasing, continuous function. The monotony arises from the additive property of arc lengths:

$$S_\alpha^t + S_t^{t+h} = S_\alpha^{t+h}.$$

Again, from the first theorem of §7, we have, for $h < \delta$,

$$S_t^{t+h} - \left| f(t+h) - f(t) \right| < \epsilon,$$

since the first term denotes an arc-length, and the second the length of a one-sided inscribed path. Therefore

$$\varphi(t+h) - \varphi(t) = S_t^{t+h} < \epsilon + \left| f(t+h) - f(t) \right|,$$

and because of the continuity of f, the right-hand side can be made as small as we please by diminishing h. Hence $\varphi(t)$ is continuous.

The continuous function

$$\psi(t) = \varphi(t) - f(t)$$

is also monotone increasing. For by the definition of φ, for $\alpha < \beta$,

$$\varphi(\beta) - \varphi(\alpha) = S_\alpha^\beta \geqslant \left| f(\beta) - f(\alpha) \right|.$$

We therefore have the desired representation:

$$f(t) = \varphi(t) - \psi(t).$$

Returning to the formula for Φ, we represent $y(t)$, which is of bounded variation, in this way:

$$y(t) = \varphi(t) - \psi(t).$$

Then
$$2\Phi = \sum_{k=1}^{n} \left[x(t_{k-1})y(t_k) - y(t_{k-1})x(t_k) \right]$$
$$= \sum_{k=1}^{n} x(t_{k-1})\left\{ y(t_k) - y(t_{k-1}) \right\} - \sum_{k=1}^{n} y(t_{k-1})\left\{ x(t_k) - x(t_{k-1}) \right\}.$$

Now
$$\sum_{k=1}^{n} x(t_{k-1})\left\{ y(t_k) - y(t_{k-1}) \right\} = \sum_{k=1}^{n} x(t_{k-1})\left\{ \varphi(t_k) - \varphi(t_{k-1}) \right\}$$
$$- \sum_{k=1}^{n} x(t_{k-1})\left\{ \psi(t_k) - \psi(t_{k-1}) \right\}.$$

Both sums on the right-hand side are in a form to which the usual existence proof for the Riemann definite integral applies. This shows that each sum:

$$\sum_{k=1}^{n} x(t_{k-1})\left\{ \varphi(t_k) - \varphi(t_{k-1}) \right\},$$

$$\sum_{k=1}^{n} x(t_{k-1})\left\{ \psi(t_k) - \psi(t_{k-1}) \right\}$$

tends to a definite limit as we refine the subdivisions of the interval $a \leqslant t \leqslant b$; that is, as we diminish δ. A similar result holds for

$$\sum_{k=1}^{n} y(t_{k-1})\left\{ x(t_k) - x(t_{k-1}) \right\},$$

as we see by representing $x(t)$ as the difference of two monotone increasing continuous functions.

Hence we finally have the theorem:

If $\epsilon > 0$ we can choose the length ρ of the side of an equilateral polygon V^* approximating to K so that between the area F enclosed by K and the area Φ^* enclosed by V^* the inequality
$$|F - \Phi^*| < \epsilon$$
holds. V^* can be chosen to have an even number of vertices.

This completes the proof of the theorems needed in §3.

Naturally the circle is a solution of many other problems in mathematics. We mention only one. If we have an enclosed plane membrane of given area, and set it vibrating, what form must the boundary take for the fundamental tone to be as deep as possible ? The answer is a circle.

It is evident that if we pursued circles conscientiously, very few branches of pure or applied mathematics would be left unexplored.

EXERCISES

1.1. With the notation of p. 1 for a triangle ABC, and the use of the power concept of p. 15, prove that $CA'.CP = CB'.CQ$, so that a circle $\mathscr{C}_3$ passes through the points A', P, B', Q. Similarly, prove that a circle $\mathscr{C}_2$ passes through the points C', R, A', P. If $\mathscr{C}_2 \neq \mathscr{C}_3$, what is their common chord? Using the theorem on radical axes of p. 21, obtain another proof of the nine-point circle theorem.

1.2. Using compasses only, determine when the construction for the inverse of a point (p. 5) breaks down. Using compasses only, show that we can find a point Y on OX such that $OY = nOX$ (integral n) (see p. 24). If n is large enough, we can find the inverse Y' of Y. Show how this enables us to find the inverse of X, when the construction for X' does not apply.

1.3. P, Q are points which are inverse in a given circle $\mathscr{C}$, and the figure is inverted with respect to any given circle Σ, so that we obtain a circle $\mathscr{C}'$ and points P' and Q'. Prove that P' and Q' are inverse points with respect to the circle $\mathscr{C}'$. What happens if $\mathscr{C}'$ is a line?

1.4. $\mathscr{C}$ and $\mathscr{D}$ are distinct circles. Find the locus of a point P which is such that two circles, each touching $\mathscr{C}$ and $\mathscr{D}$, also touch at P. (There are three cases to examine: $\mathscr{C}$ intersects $\mathscr{D}$, $\mathscr{C}$ touches $\mathscr{D}$, and $\mathscr{C}$ does not intersect $\mathscr{D}$.)

1.5. Prove that two circles $\mathscr{C}$ and $\mathscr{D}$ which intersect at B and C are orthogonal if and only if there exist two circles touching $\mathscr{D}$ at B and at C respectively which also touch each other at A, where A is any point on $\mathscr{C}$ distinct from B and C.

1.6. Let $\mathscr{C}$ and $\mathscr{D}$ be circles which intersect in the points A and B. By inverting with A as centre, show that the operations of inversion in $\mathscr{C}$ and $\mathscr{D}$ are commutative if and only if $\mathscr{C}$ and $\mathscr{D}$ are orthogonal. (If inversion in $\mathscr{C}$ is represented by $T_{\mathscr{C}}$, and inversion in $\mathscr{D}$ is represented by $T_{\mathscr{D}}$, we wish to prove that if P is any point then $(P)T_{\mathscr{C}}T_{\mathscr{D}} = (P)T_{\mathscr{D}}T_{\mathscr{C}}$ if and only if $\mathscr{C}$ and $\mathscr{D}$ are orthogonal. The theorem of Exercise 1.3 is to be used.)

1.7. In Exercise 1.6, above, can inversion be commutative if $\mathscr{C}$ and $\mathscr{D}$ touch, or if $\mathscr{C}$ and $\mathscr{D}$ do not intersect?

1.8. In triangle ABC, $\angle BAC = \alpha > 120°$. P is any point inside the triangle distinct from A. Let $\beta = 180° - \alpha$, so that $\beta < 60°$. Rotate triangle ABP about the vertex A through $\beta°$ to triangle $AB'P'$, A remaining fixed, B mapping on B' and P on P'. Then $\angle BAB' = \beta$, and since $\beta + \angle BAC = 180°$, the points B', A, C are collinear. Prove that in the isosceles triangle PAP' we have $AP > PP'$, and that

$$AP + PB + PC > PP' + P'B' + PC = B'P + P'P + PC$$
$$\geqslant B'C = B'A + AC = AB + AC,$$

so that A is the Fermat point in this case (p. 12). (This proof is due to Dan Sokolowsky.)

1.9. Show that the power of a point P with respect to a point-circle (one of zero radius) centre A is equal to $(PA)^2$. Hence determine the locus of a point P which moves so that the ratio $PA : PB$ of its distances from two given points A and B is constant.

1.10. A is a point of intersection of circles $\mathscr{C}$ and $\mathscr{D}$, and a line through A intersects $\mathscr{C}$ again in P, and $\mathscr{D}$ again in Q. If V is the midpoint of PQ, prove that as the line through A varies, the point V moves on a circle. (Compare the powers of V with respect to $\mathscr{C}$ and to $\mathscr{D}$.) Examine the case when V divides PQ in a given ratio.

1.11. QR is a chord of a circle $\mathscr{C}$ which subtends a right angle at a given point L (the angle QLR is a right angle). If P is the midpoint of QR, prove that the power of P with respect to $\mathscr{C}$ is $-(PL)^2$. Deduce that the locus of P as the chord QR varies is a circle in the pencil determined by $\mathscr{C}$ and the point-circle L.

1.12. Prove that the coaxal system of circles $\lambda C + \mu C' = 0$, where $C \equiv 3(x^2 + y^2) - 2x - 4y + 3 = 0$, $C' \equiv 3(x^2 + y^2) - 4x - 2y + 3 = 0$ contains two point-circles, and that the equation of the coaxal system may be written in the form:

$$k_1[(x - 1)^2 + y^2] + k_2[x^2 + (y - 1)^2] = 0.$$

1.13. If $\mathscr{C}$ and $\mathscr{C}'$ are non-intersecting circles, and $\mathscr{C}_1, \mathscr{C}_2, \mathscr{C}_3, \ldots, \mathscr{C}_n$ touch $\mathscr{C}$ and $\mathscr{C}'$ and also each other as in Fig. 25 (p. 20), prove that the points of contact of the circles $\mathscr{C}_i$ with each other lie on a circle $\mathscr{D}$, and that the inverse of $\mathscr{C}$ in $\mathscr{D}$ is the circle $\mathscr{C}'$.

1.14. $\mathscr{C}_1$ and $\mathscr{C}_2$ are point-circles, and $\mathscr{C}_3$ is a proper circle. Prove that there are two circles which pass through the point-circles and touch $\mathscr{C}_3$.

1.15. $\mathscr{C}_1$ is a point-circle, $\mathscr{C}_2$ and $\mathscr{C}_3$ are proper circles which touch each other.   Prove that there are three solutions to the Apollonian problem.

1.16. $\mathscr{C}_1$, $\mathscr{C}_2$ and $\mathscr{C}_3$ are proper circles, and $\mathscr{C}_1$ touches both $\mathscr{C}_2$ and $\mathscr{C}_3$, but these two do not touch each other.   Prove that there are four solutions to the Apollonian problem, and five if we include $\mathscr{C}_1$.

1.17. $\mathscr{C}_1$ and $\mathscr{C}_2$ touch, and $\mathscr{C}_3$ is a proper circle which does not touch either.   Prove that there are six distinct circles which touch the three given circles.

1.18. Using compasses only, justify the following construction for determining the centre of a given circle $\mathscr{C}$. With centre at any point A of $\mathscr{C}$, draw a circle $\mathscr{D}$ to cut $\mathscr{C}$ at points B and C.   Find the geometric image O' of A in the line BC.   Then the inverse of O' in the circle $\mathscr{D}$ is the centre of the circle $\mathscr{C}$.

1.19. A triangular cut-out ABC is given and we move it so that the side AB always passes through a given point L, and the side AC always passes through another given point M.   Prove that the side BC of the cut-out touches a fixed circle.

1.20. *The Bobillier Envelope Theorem.*   A triangular cut-out ABC is given, and we move it so that AB always touches a given circle $\mathscr{C}$, and AC always touches another given circle $\mathscr{D}$.   Prove that the side BC of the cut-out touches a fixed circle $\mathscr{E}$.

CHAPTER II

2.1. Show that a line l which lies in the Oxy-plane in E_3 represents a coaxal system of circles which all pass through the point $(0,0)$.   Where is the other common point of intersection?   What kind of coaxal system is represented by l if l passes through $(0,0)$ and lies in the Oxy-plane?

2.2. Show that the coaxal system determined by the two circles:

$$C \equiv x^2 + y^2 - 2px - 2qy - k^2 = 0,$$
$$C' \equiv x^2 + y^2 - 2p'x - 2q'y - k^2 = 0$$

is always of the intersecting type, with distinct points of intersection if $k \neq 0$.

2.3. Show that two coaxal systems of circles in the Oxy-plane whose maps in E_3 are parallel lines are coaxal systems with the same radical axis.

2.4. Verify that for all a, b, and $c \neq 0$, every circle of the system:

$$a(x - x') + b(y - y') + c(x^2 + y^2 - z') = 0$$

intersects the fixed circle of the system:

$$-2x'(x - x') - 2y'(y - y') + x^2 + y^2 - z' = 0$$

at the ends of a diameter of this fixed circle.

2.5. A plane in E_3 which contains the point (p,q,r) has the equation:

$$a(x - p) + b(y - q) + c(z - r) = 0.$$

Deduce that the circle $a(x - p) + b(y - q) + c(x^2 + y^2 - r) = 0$ is always orthogonal to the circle $x^2 + y^2 - 2px - 2qy + r = 0$.

2.6. Prove that the circle which is orthogonal to each of three given circles $\mathscr{C}_1$, $\mathscr{C}_2$, $\mathscr{C}_3$ is uniquely defined, unless the three given circles belong to a coaxal system. If this is the case, are there any circles which touch the three given circles?

2.7. If three given circles are not in a coaxal system, and $\mathscr{C}_0$ is the uniquely defined common orthogonal circle (Exercise 2.6, above), show that the eight Apollonian contact circles (p. 21) are inverse in pairs in $\mathscr{C}_0$.

Chapter III

3.1. Prove that any Möbius transformation with the two distinct fixed points p and q may be written in the form:

$$(w - p)/(w - q) = k(z - p)/(z - q),$$

where k is a complex number. (A fixed point under a Möbius transformation is one which is mapped onto itself.)

3.2. Show that the points of any given circle in the z-plane may be represented thus: $z = (At + B)/(Ct + D)$, where A, B, C, D are complex numbers with $AD - BC \neq 0$, and t is real.

3.3. If ABC is an h-triangle in the Poincaré model (p. 58), where A is at the centre of ω, so that AB and AC are diameters of ω, show that BC is convex to A, and that the sum of the angles of the h-triangle ABC is less than the sum of the angles of the Euclidean triangle ABC, which is 180°.

3.4. If ABC is any h-triangle in Ω, prove that the angle sum is less than 180°. (Find an inversion which maps ω into itself and A into the centre.)

3.5. Show that through any given h-point A there exists a unique h-line l' perpendicular to a given h-line l, and that if m, n are the h-lines through A which are h-parallel to l, it being assumed that A does not lie on l, then m and n make equal angles with l'.

3.6. Prove that two distinct h-lines which are each perpendicular to the same h-line have no intersection.

3.7. Prove that two distinct h-lines which have no intersection have a unique h-line perpendicular to both of them.

SOLUTIONS

Chapter I

1.1. Circle on AB as diameter through P, Q. Therefore $CP.CB = CQ.CA$, but $CB = 2CA'$, $CA = 2CB'$. Common chord of $\mathscr{C}_2$, $\mathscr{C}_3$ is BC. Common chords of three circles cannot form sides of triangle. Hence one circle through A', P, R, C', Q, B'. Apply this theorem to triangle BHC, which shows above circle also goes through midpoints of HA, HB and therefore of HC.

1.2. If $OX < \frac{1}{2}k$, choose n so that $OY = nOX > \frac{1}{2}k$, and find Y', where $OY.OY' = k^2 = (nOX)(OY')$. Find X', where $X' = nOY'$, then $OX.OX' = k^2$.

1.3. All circles $\mathscr{D}$ through P, Q are orthogonal to $\mathscr{C}$. Therefore all circles $\mathscr{D}'$ through P', Q' are orthogonal to $\mathscr{C}'$. Hence P', Q' inverse in $\mathscr{C}'$. If $\mathscr{C}'$ is a line, P', Q' are mirror images in line.

1.4. If $\mathscr{C}$, $\mathscr{D}$ intersect at A, B, take A as centre of inversion. Locus of P' consists of two angle-bisectors of lines $\mathscr{C}'$, $\mathscr{D}'$ intersecting at B', so locus of P consists of two circles through A, B. If $\mathscr{C}$, $\mathscr{D}$ touch at A, $\mathscr{C}'$, $\mathscr{D}'$ are parallel lines, P' moves on halfway parallel line, and P moves on circle which touches both $\mathscr{C}$ and $\mathscr{D}$ at A. If $\mathscr{C}$, $\mathscr{D}$ do not intersect, invert into concentric circles, and P' moves on a third concentric circle, and therefore P moves on a circle in coaxal system determined by $\mathscr{C}$ and $\mathscr{D}$.

1.5. Invert, centre B, then $\mathscr{C}'$, $\mathscr{D}'$ are lines intersecting at C'. Circle touching $\mathscr{D}$ at B inverts into line parallel to $\mathscr{D}'$, intersecting $\mathscr{C}'$ at A'. Circle touching $\mathscr{D}$ at C inverts into circle touching $\mathscr{D}'$ at C' and through A'. The line through A' parallel to $\mathscr{D}'$ touches this latter circle at A' if and only if $\mathscr{C}'$ is orthogonal to $\mathscr{D}'$.

1.6. Take A as centre of inversion, then $\mathscr{C}'$ and $\mathscr{D}'$ are lines intersecting at B', and inversions in $\mathscr{C}$ and $\mathscr{D}$ become line reflections in $\mathscr{C}'$ and $\mathscr{D}'$. These are commutative if and only if $\mathscr{C}'$ is perpendicular to $\mathscr{D}'$, since a composition of reflections in two intersecting lines is equivalent to a rotation about their intersection through twice the angle between the lines, measured from the line of the first reflection to the second line.

1.7. No! Composition of reflections in parallel lines (after inversion) is equivalent to a translation, directed from the first line to the second, and composition of inversions in concentric circles (after inversion of the two non-intersecting circles) is equivalent to an enlargement with scale-factor depending on the order of inversion.

1.8. In a triangle the greatest side is opposite the greatest angle, and $\angle PP'B > 60° > \beta$. The shortest distance from B' to C is $= B'C$.

1.9. The equation to a point-circle (α,β) is $(x - \alpha)^2 + (y - \beta)^2 = 0$, and the power of $P(x,y)$ is $(x - \alpha)^2 + (y - \beta)^2 = (PA)^2$. Locus is circle of coaxal system determined by point-circles A and B (p. 15).

1.10. Power of V with respect to $\mathscr{C} = VA.VP$, and with respect to $\mathscr{D} = VA.VQ$. If $VP = -VQ$, then ratio of powers $= -1$, so by theorem on p. 15 the point V moves on circle through intersections of $\mathscr{C}$ and $\mathscr{D}$. Similar result if V divides PQ in a given ratio.

1.11. $PQ = PR = PL$, and so power of P with respect to $\mathscr{C} = -(PQ)^2 = -(PL)^2$. Power of P with respect to point-circle $L = (PL)^2$, so ratio of powers $= -1$, and theorem on p. 15 gives result.

1.12. Find the radius of $\lambda C + \mu C' = 0$, and equate to zero, obtaining a quadratic equation giving point-circles $(x - 1)^2 + y^2 = 0$, $x^2 + (y - 1)^2 = 0$, and use result at bottom of p. 15.

1.13. Invert $\mathscr{C}$, $\mathscr{C}'$ into concentric circles $\mathscr{E}$, $\mathscr{E}'$, centre V, and suppose that $\mathscr{C}_i$, $\mathscr{C}_{i+1}$ invert into $\mathscr{E}_i$, $\mathscr{E}_{i+1}$ which touch each other at T, $\mathscr{E}_i$ touches $\mathscr{E}$ at A and $\mathscr{E}'$ at A', and $\mathscr{E}_{i+1}$ touches $\mathscr{E}$ at B and $\mathscr{E}'$ at B'. Then tangent at T to $\mathscr{E}_i$, $\mathscr{E}_{i+1}$ passes through V, as do the lines AA', BB', and $(VT)^2 = (VA)(VA') = (VB)(VB')$.

1.14. Take point $\mathscr{C}_1$ as centre of inversion. There are two tangents from point $\mathscr{C}_2'$ to circle $\mathscr{C}_3'$, and these arise from the circles sought.

1.15. Take point of contact of $\mathscr{C}_2$, $\mathscr{C}_3$ as centre of inversion, then $\mathscr{C}_2'$, $\mathscr{C}_3'$ are parallel lines, and we seek circles through point $\mathscr{C}_1'$ which touch these lines. One such circle is the line through $\mathscr{C}_1'$ parallel to $\mathscr{C}_2'$, and this arises from solution touching $\mathscr{C}_2$, $\mathscr{C}_3$ at point of contact.

1.16. Take point of contact of $\mathscr{C}_1$ and $\mathscr{C}_3$ as centre of inversion. Then $\mathscr{C}_1'$ and $\mathscr{C}_3'$ are parallel lines, and $\mathscr{C}_2'$ is a circle touching $\mathscr{C}_1'$. One solution is the line parallel to $\mathscr{C}_1'$ which touches $\mathscr{C}_2'$, and there are three circles which touch $\mathscr{C}_1'$, $\mathscr{C}_2'$ and $\mathscr{C}_3'$.

1.17. Take point of contact of $\mathscr{C}_1$ and $\mathscr{C}_2$ as centre of inversion. $\mathscr{C}_1'$ and $\mathscr{C}_2'$ are parallel lines, and $\mathscr{C}_3'$ is a circle which touches neither. Two solutions are tangents to $\mathscr{C}_3'$ which are parallel to $\mathscr{C}_1'$.

1.18. Invert in $\mathscr{D}$. Then circle $\mathscr{C}$ becomes the line BC, and centre of $\mathscr{C}$ inverts into the point O' which is the inverse of the centre of $\mathscr{D}$, the point A, in the inverse of $\mathscr{C}$, which is the line BC. The inverse of O' in $\mathscr{D}$ is the centre O of the circle $\mathscr{C}$.

1.19. The circle LAM remains fixed as A moves. If AU is parallel to BC, meeting this circle again in U, show that U is fixed on this circle. It is the centre of the envelope of BC.

1.20. Reduce this case to the one treated above by drawing a parallel to AB to pass through the centre of $\mathscr{C}$, and a parallel to AC to pass through the centre of $\mathscr{D}$.

Chapter II

2.1. The circle $x^2 + y^2 - 2px - 2qy + r = 0$ is mapped on (p,q,r). If $r = 0$ the circle passes through $(0,0)$. The centres (p,q) of the circles lie on l, and the reflection of $(0,0)$ in the line l is the other common point of intersection. If l passes through $(0,0)$, the circles all touch at $(0,0)$.

2.2. The two circles are mapped onto the points $(p,q,-k^2)$, $(p',q',-k^2)$, and the line joining these points is parallel to the Oxy-plane and below Ω, so that if $k \neq 0$ it does not intersect Ω. Hence its polar line intersects Ω in two distinct points, and the coaxal system is of the intersecting type, with two distinct points of intersection.

2.3. If the parallel lines in E_3 are $(x - a)/l = (y - b)/m = (z - c)/n = t$, and $(x - a')/l = (y - b')/m = (z - c')/n = t'$, the circles are

$$x^2 + y^2 - 2(a + lt)x - 2(b + mt)y + c + nt = 0,$$
$$x^2 + y^2 - 2(a' + lt')x - 2(b' + mt')y + c' + nt' = 0,$$

for varying t and t', and each system has the radical axis:

$$2lx + 2my - n = 0.$$

2.4. Common chord of the two circles is:

$$(x - x')(a/c + 2x') + (y - y')(b/c + 2y') = 0.$$

2.5. Circles are orthogonal if their representative points in E_3 are conjugate. The point (p,q,r) is in the given plane, and therefore conjugate to the pole of the given plane.

2.6. Three points in E_3 which are not collinear define a plane with a unique pole with respect to Ω. The pole represents $\mathscr{C}_0$, the unique circle orthogonal to the three given circles. If the three circles are in a coaxal system of the intersecting type, the point-circles of intersection are the only contact circles. If non-intersecting, there are no contact circles.

2.7. Since $\mathscr{C}_i$ is orthogonal to $\mathscr{C}_0$, its inverse in $\mathscr{C}_0$ is itself. If we invert in $\mathscr{C}_0$ the three given circles invert into themselves, and contact circles invert into contact circles, so that the eight contact circles are paired.

CHAPTER III

3.1. The equation of the transformation which has (a,a'), (p,p), (q,q), and (z,w) as corresponding pairs of points is given by (2), p. 50, as:

$$(wa',pq) = (za,pq)$$

which leads immediately to the required form.

3.2. Since the cross-ratio of four points on a circle is real, we have $(za,bc) = t$, where a, b and c are fixed points on the circle, z a variable point, and t is real. From $(z - b)/(z - c) = t(a - b)/(a - c)$, we find z in the required form. If $AD - BC = 0$, we should find that $At + B = k(Ct + D)$, so that $z = k$, whereas z is a variable point on the circle.

3.3. Since the circular arc BC is orthogonal to ω the centre of the arc lies outside Ω, and therefore the arc BC is convex to the centre of ω.

3.4. The arcs AB and AC both pass through the inverse A' of A in ω. Take A' as centre of inversion, and for circle of inversion the circle Σ centre A' orthogonal to ω. Then ω inverts into itself, and since circles through A' and A invert into lines orthogonal to ω through the inverse of A in Σ, this inverse is the centre of ω.

3.5. If the h-line intersects ω in B and C, after an inversion which maps ω onto itself, B onto B' and C onto C', and A onto the centre O of ω, the inverse of the h-line BC is the h-line $B'C'$, and the Euclidean line through O and the centre Q of the circle defined by the arc $B'C'$ is perpendicular to the h-line $B'C'$, the lines joining the point O to B' and C' respectively touch the h-line $B'C'$, and OQ bisects the angle $B'OC'$.

3.6. If the given h-line is l, and the perpendicular h-lines are m and n, l and ω define a coaxal system of the intersecting type, and m and n are circles in the conjugate coaxal system, which is non-intersecting.

3.7. Two distinct h-lines with no intersection define a non-intersecting coaxal system, and have limiting points L and L' which lie on ω, since this is a member of the conjugate coaxal system. There is a unique circle through L and L' which is orthogonal to ω, and this is the unique h-line which is perpendicular to the given h-lines.

APPENDIX

Karl Wilhelm Feuerbach, Mathematician

by Laura Guggenbuhl*

One of the most important theorems of the geometry of the triangle that was developed during the nineteenth century says that "the nine point circle of a triangle is tangent to the inscribed and to each of the three escribed circles of the triangle." The theorem was discovered and first proved by Karl Wilhelm Feuerbach, a little-known mathematician who led a short and tempestuous life. Although modern textbooks in college geometry discuss the theorem at great length, a search for additional details has been extremely interesting and fruitful (**1**). Furthermore, it seems that few people realize that Feuerbach, the mathematician, was a member of the famous Feuerbach family of Germany.

An early date in this story is 1821. It was in this year that Brianchon and Poncelet proved (**2**) that

... the circle through the feet of the altitudes of a triangle also passes through the midpoints of the sides, as well as the midpoints of the segments of the altitudes from their point of intersection to the vertices of the triangle.

Brianchon and Poncelet did not seem to have any idea that the circle in question would attract the interest of so many mathematicians in years to come. They did not give a name to the circle, and after a page or two they went on to another problem.

At almost the same time, a slender little volume of a bare 62 pages appeared in Nuremburg under the formidable title *Eigenschaften einiger merkwurdigen Punkte des geradlinigen Dreiecks und mehrerer durch sie bestimmten Linien und Figuren. Eine analytisch-trigonometrische Abhandlung* (Properties of some notable points of a plane triangle, and various lines and figures

---

* Dr. Guggenbuhl was assistant professor of mathematics at Hunter College in New York. She received her training at Hunter College and Bryn Mawr College. This article is based on a paper presented by the author before Section L, History and Philosophy of Science, at the Boston Meeting of the AAAS, December 1953.

Reprinted with permission from *The Scientific Monthly,* Vol. 81 (1955) pp. 71–76.

determined by these points, an analytic-trigonometric treatise). The title page of this book describes the author as Karl Wilhelm Feuerbach, doctor of philosophy. It may be that the book was Feuerbach's doctoral dissertation, but it is not possible to be certain whether this was the case or not. Feuerbach attended several different schools and universities, but a careful search has failed to uncover the title of his dissertation or the name of the university that awarded him the doctor's degree. At any rate, with the appearance of this book in 1822, Feuerbach completed his college studies at the University of Freiburg (Br.).

This book probably would have been long since forgotten had it not been for an inconspicuous theorem at the bottom of page 38, which states simply,

> ... the circle which passes through the feet of the altitudes of a triangle is tangent to all four circles, which in turn are tangent to the three sides of the triangle. ...

Feuerbach's proof was purely algebraic and admittedly modeled upon the proof in an article by Euler (3) of 1765. Feuerbach showed that the two circles in question were tangent by proving that the distance between the centers was equal to the sum of the radii of the circles. In outline, the proof sounds simple enough. However, one can appreciate the skill and patience that led to this discovery only by working out the details of the algebraic identities involved in the development.

Feuerbach gives only three or four references in addition to the article by Euler. It is particularly interesting to note that he refers to an article by L'Huillier, dated 1810, which appeared in Gergonne's *Annales de mathematiques,* and thus indicated that he was familiar with the journal that carried the aforementioned article by Brianchon and Poncelet. However, Feuerbach makes no reference to the work of Brianchon and Poncelet and was probably unaware of its existence.

One can scarcely say, as some writers claim, that Feuerbach discovered the nine-point circle of a triangle. For at no place in his book does he mention the three points that are the midpoints of the segments of the altitudes from the orthocenter to the vertices of the triangle, and he does not letter these points on his diagram. Feuerbach's description of the circle that was to become so closely associated with his name was merely the unimagina-

VI. *Inueſtigatio diſtantiae punctorum* G *et* H.

114   S O L V T I O

17. Pro hoc caſu poſtremo habetur :

$$AR - AS = \frac{c+b-a}{2} - \frac{a}{2}; c = \frac{b-a}{2}$$

$$RG - SH = \frac{2A}{a+b+c} - \frac{c'aa + bb \cdot cc)}{2A} = \frac{(a+b) \cdot{}^2 + (aa + bb)cc - (c+)'(a+b)k - (aa - bb)k}{2(a+b+c)A}$$

quarum binarum formularum quadrata ſi addantur reperitur ſequens expreſſio :

$$GH' = \frac{abc}{16(a+b+c)^2 AA} \left\{ \begin{array}{l} +a' + a'b + ab' + abc' - 2a'bb - 2aab' \\ + b' + a'c + ac' + ab'c - 2a'cc - 2aac' \\ + c' + b'c + bc' + a'bc - 2b'cc - 2bbc' \end{array} \right\}$$

quae per $a + b + c$ reducta abit in hanc :

$$GH' = \frac{abc}{16(a+b+c)AA} \left\{ \begin{array}{l} +a' + aabc - 2aabb \\ + b' + abbc - 2aacc \\ + c' + abcc - 2bbcc \end{array} \right\}$$

Vnde facta ſubſtitutione colligitur

$$GH' = \frac{r}{16 \, pAA}(p' - 4ppq + 9pr) - \frac{r(p' - 4pq + 9r)}{16 \, AA}$$

ſeu $GH' = \frac{rr}{16 \, AA} - \frac{r}{p}$.

Tom. XI. Nou. Comm.   P   18. En

18. En ergo ſub vno conſpectu quadrata ſex horum interuallorum :

  I. $EF' = \frac{rr}{16 \, AA} - \frac{1}{4}(pp - 2q)$

  II. $EG' = \frac{rr}{16 \, AA} - pp + 3q - \frac{2r}{p}$

 III. $EH' = \frac{9rr}{16 \, AA} - pp + 2q$

 IV. $FG' = -\frac{1}{2}pp + \frac{1}{2}q - \frac{1}{2}\frac{r}{p}$

  V. $FH' = \frac{rr}{16 \, AA} - \frac{1}{4}(pp - 2q)$

 VI. $GH' = \frac{rr}{16 \, AA} - \frac{r}{p}$

Portions of pages 113 and 114 of Euler's article "Solutio Facilis Problematum Quorumdam Geometricorum Difficillimorum." Here we see the development of a series of algebraic identities concerning the sides of a triangle. [Zentralbibliothek, Zurich]

tive phrase "the circle which passes through the points M, N, and P (the feet of the altitudes of the triangle)." It is nevertheless a fact that in Europe the circle is usually called the circle of Feuerbach. It seems too that Feuerbach did not realize that this one theorem was so much more significant than the others with which it was listed, and he surely did not realize that his fame as a mathematician would rest upon this single theorem.

Shortly after the publication of his book, Feuerbach, only 22 years old and without previous teaching experience, was named professor of mathematics at the Gymnasium at Erlangen. There is nothing else to indicate that the book made a deep impression upon contemporary mathematicians, or even that it had a wide circulation. In fact, 11 years later (in 1833) Steiner, writing from Berlin, said he did not know that the theorem in question had been previously proved by Feuerbach (4). In another note at the end of his book, Steiner said that the "circle is now generally known as the circle of Feuerbach." In 1842 the theorem was again submitted as an original contribution to mathematical literature and proved anew by Terquem (5). It is in this article by Terquem that the circle is first designed as the nine-point circle.

In succeeding years many articles were written about the nine-point circle. People seemed to be fascinated by the difficulties in Feuerbach's proof and seemed to find it a pleasant chal-

lenge to produce a different and far simpler proof. Two particularly detailed and exhaustive treatments appeared almost simultaneously near the end of the 19th century, one an article by
the well-known Mackay (6), and the other an article by Julius
Lange, professor of mathematics at the Friedrichs Wederschen
Ober-Realschule in Berlin (7).

It is a little strange that so much has been written about
the theorem Feuerbach discovered and that so little has been
written about the man who discovered the theorem. In an article
by Moritz Cantor (8) a few pages are devoted to biographical
details, but by far the most poignant story of Feuerbach's short
life can be read in his father's letters (9).

Karl Wilhelm was the third son in a family of 11 children
born to the famous German jurist Paul Johann Anselm Feuerbach, and Eva Wilhelmine Maria Troster. The first eight children in the family were boys, and the last three were girls. Since
three of the boys died in infancy, Karl is usually spoken of as
the second of five sons of the jurist. These five sons were Joseph
Anselm (1798–1851) philologist and archeologist, and father of

$$\longrightarrow \quad 35 \quad \longrightarrow$$

$$\overline{OS}^2 = (AF - AP)^2 + (OP - SF)^2.$$

Nun ift aber $AF = \frac{1}{2}(-a+b+c)$ und $AP = \dfrac{-a^2+b^2+c^2}{2c}$, folglich:

$$AF - AP = \frac{(a-b+c)(a+b-c) - c(a+b-c)}{2c};$$

ferner, weil (§. 55.) $OP = \dfrac{(-a^2+b^2+c^2)(a^2-b^2+c^2)}{8c\triangle}$, und (§. 2.) $SF = \dfrac{2\triangle}{a+b+c}$,

fo ift:

$$OP - SF = \frac{(-a^2+b^2+c^2)(a^2-b^2+c^2) - c(-a+b+c)(a-b+c)(a+b-c)}{8c\triangle}.$$

Subſtituirt man nun im Ausdruce für $\overline{OS}^2$, fo wird man denſelben endlich in dieſe Form bringen können:

$$\overline{OS}^2 = \frac{(-a+b+c)^2(a-b+c)^2(a+b-c)^2 - (-a^2+b^2+c^2)(a^2-b^2+c^2)(a^2+b^2-c^2)}{32\triangle^2},$$

woraus ſich durch Einführung der Kreishalbmeſſer r, ρ, R ergiebt:

$$\overline{OS}^2 = 2r^2 - 2\rho R$$

Algebraic identities on page 35 of Feuerbach's *Eigenschaften einiger
merkwurdigen Punkte. ...* Note the similarity between these algebraic
identities and those on pages 113 and 114 of Euler's "Solution Facilis ... ."
[New York Public Library]

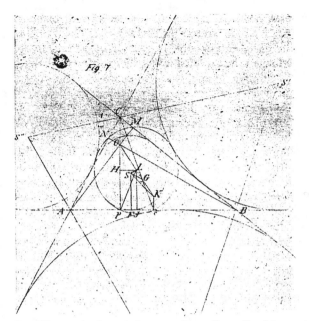

Diagram used in *Eigenschaften einiger merkwurdigen Punkte* ... to illustrate
the theorem of Feuerbach. Note that three of the famous nine points,
namely, the midpoints of the segments of the altitudes from the orthocenter
to the vertices of the triangle, are not named on this diagram. [New York
Public Library]

the famous German painter Anselm Friedrich (1829–1880); Karl
Wilhelm (1800–1834), mathematician, the subject of this article;
Eduard August (1803–1843), professor of law at the University
of Erlangen; Ludwig Andreas (1804–1872), famous philosopher;
and Friedrich Heinrich (1806–1880), orientalist.

Karl was born in Jena 30 May 1800. His father, 25 years old
when Karl was born, had left his home in Frankfurt at the age of
16 to study philosophy at the University of Jena. However, since
there were jurists and public servants on both sides of his family,
it is not surprising to learn that, as his family responsibilities
increased, he was soon attracted to the more lucrative profession
of the law. All manner of success came to him at a very early
age. But it was a number of years before his own father, who
had been angered when he left home, became reconciled to him.
The young man sent many entreating letters to his father that
remained unanswered. It is in one such ingratiating letter that
one can read the earliest reference to Karl; his father says, in a

letter from Kiel dated 16 November 1802, that "his youngest son
Karl is a healthy, red cheeked, fat youngster, who runs around
happily and has little thought for anything except food." In 1804
the father writes of taking his wife and three young sons with
him from Kiel to Landshut in an open wagon in freezing winter
weather.

As increasing fame and success came to the father, he moved
frequently from one city to another in Germany. Thus Karl, while
still a young boy, lived successively in Jena, Kiel, Landshut, and
Munich. When Karl was 14 years old his father moved again from
Munich to Bamberg. At this time, however, he left his two eldest
sons, Anselm and Karl, in school in Munich. In 1817 Anselm
and Karl entered the University of Erlangen. At Erlangen they
studied under the patronage of King Maximilian Joseph, who had
ennobled their father, and who had promised to provide for the
university education of all the jurist's sons.

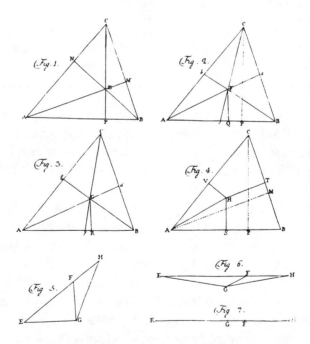

Diagrams used to illustrate Euler's article "Solutio Facilis Problematum
Quorumdam Geometricorum Difficillimorum." In these diagrams, the point
E is the orthocenter of triangle *ABC*; *F* the median point; *G* the incenter;
and *H* the circumcenter. In Figs. 5 and 6 the line *EFH* is the Euler line of
the triangle. [Zentralbibliothek, Zurich]

Karl Wilhelm Feuerbach. Published for the first time in *Genie und Krankheit* by Theodor Spoerri. [New York Academy of Medicine]

By 1819 the father had become president of the court of appeals in Ansbach. In an 1819 Christmas letter from Ansbach to his father, the jurist writes at great length about Karl's gifts in mathematics and physics. He says that Karl had given some thought to jurisprudence as a profession, but that it was his present ambition to become an engineer in the army.

After a short stay at Erlangen, Karl transferred to the University of Freiburg (Br.). He did so in order that he might continue his studies in mathematics with Karl Buzengeiger, who is credited with having had a great influence upon the young mathematician. In fact, Feuerbach's 1822 publication *Eigenschaften einiger merkwurdigen Punkte* ...) features an introduction (of 16 pages) by this same professor. No doubt Buzengeiger wished to use this means to introduce a talented young student to the academic world. However, in his introduction, he says scarcely a word about the student or any of the geometric discoveries that followed in the ensuing pages.

As the time passes, one can read the gradually growing pride and joy of the father in the son's accomplishments. He writes of the little book that Karl published in 1822 and asks a dear friend to do what she can to bring the book to the attention of influential people in Bavaria. The Feuerbachs were, after all, not native Bavarians; and moreover they were Protestants in a predominantly Catholic state. For these reasons the father feared that the son might encounter some difficulty in finding a place in the academic profession. When the son was appointed professor of mathematics at the Gymnasium at Erlangen everything seemed to be going well.

And then very suddenly, completely without warning and in a most humiliating manner, the father's pride, ambitions, and joy in the son Karl were shattered by a devastating blow. One day, as Karl was walking to school, he was arrested on the street. He was one of a group of 20 young men who were rounded up and later imprisoned in the New Tower in Munich. The men were arrested because of the political nature of the activities of an organization to which they had belonged as undergraduates. Karl had been known as a gay and carefree student in the early days of his university career. He had incurred debts, and he was at times impetuous and unconventional. However, between the lines of his father's letters, one can read the thought that the incident may have been instigated by the father's political enemies.

In a long letter of 1 July 1827, the father pours out the anguish in his heart to a very dear personal friend. He tells how the weeks and months were allowed to drag on without action and that Karl was not allowed to receive letters or visitors. Karl became obsessed by the idea that only his death could free his 19 companions. He made two separate attempts at suicide. One evening he was found unconscious from loss of blood after having cut the veins in his feet. He was transferred to a hospital and then made a second attempt at suicide by jumping out a window. He was saved from death by landing in a deep snow bank but was permanently crippled as a result of the accident. The father writes bitterly of the fact that he was not more closely guarded, particularly since he had been declared on the verge of a mental breakdown when he was arrested. A short time later, Karl was paroled in the custody of his former teacher in Munich, Friedrich Thiersch (**10**), a close friend of the Feuerbach family. One of the 20 young men died while in prison. Finally, after 14 months, a trial was held, the men were vindicated, released, and allowed to resume their normal activities.

For about a year Karl seemed to want nothing more than the companionship of his brothers and sisters, the comfort of his father's home at Ansbach, and the peace and quiet that would allow him to complete some mathematical research on which he had been working while he was in prison in Munich. It is doubtful that this piece of work could have had any greater compliment than the critical analysis to which it was subjected by Cantor in his article in the *Sitzungsberichte der Heidelberg Akademie der Wissenschaften* (**8**). In a final summary Cantor says that Feuerbach in this work proved himself to be an independent codiscoverer with Möbius of the theory of the homogeneous coordinates of a point in space.

Feuerbach's development of the theory of homogeneous coordinates is his second and last study. It is contained in three separate units. First there was a brief notice in Oken's *Isis* (**11**) written from Ansbach and dated 22 October 1826 under the heading "Einleitung zu dem Werke Analysis der dreyeckigen Pyramide durch die Methode der Coordinaten und Projectionen. Ein Beytrag zu der analytischen Geometrie von Dr. Karl Wilhelm Feuerbach, Prof. d. Math." (Introduction to the analysis of the triangular pyramid, by means of the methods of coordinates and projections. A study in analytic geometry). In this brief notice, Feuerbach gives a summary of his results, lists references, and says that he hopes to find a publisher for the entire work. The following year, 1827, the material was published in a 48-page booklet at Nuremburg under the title *Grundriss zu analytischen Untersuchungen der dreyeckigen Pyramide* (Foundations of the analytic theory of the triangular pyramid). The third unit in this study is an unpublished manuscript in Feuerbach's handwriting, dated 7 July 1826—a manuscript that is discussed at great length by Cantor (**8**). However, Feuerbach's work on the triangular pyramid did not capture the imagination, as did his earlier publication *Eigenschaften einiger merkwurdigen Punkte.* . . . Many years after his death an effort was made to find a publisher for his manuscript of 1826, but these efforts were unsuccessful. In passing it is worthy of note that Feuerbach's *Grundriss.* . . , published at

Nuremburg, and Möbius' *Der barycentrische Calcul,* published at
Leipzig, both bear the date 1827.

When the young men who were arrested with Feuerbach had
finally been released from prison, King Maximilian Joseph took
great pains to help them return to a normal life. Feuerbach was
appointed professor of mathematics at the Gymnasium at Hof.
But at Hof Feuerbach was far from happy. Apparently he found
no substitute at Hof for the companionship of his brothers and
sisters, for the vivid personalities in his father's circle of friends,
and for the gay university life at Erlangen. Before long he suf-
fered such a severe relapse in his illness that he was forced to
give up his teaching. Two of his brothers, Eduard and Ludwig,
brought him back from Hof to Erlangen for medical treatment.
By 1828 he had recovered sufficiently to resume teaching at the
Gymnasium at Erlangen. But one day he appeared in class with a
drawn sword and threatened to cut off the head of every student
in the class who could not solve the equations he had written on
the blackboard. After this episode he was permanently retired.
Gradually he withdrew more and more from reality. He allowed
his hair, beard, and nails to grow long; he would stare at occa-
sional visitors without any sign of emotion; and his conversation
consisted only of low mumbled tones without meaning or expression. And then there is no further word about Karl, not even in his
father's letters. His father ends a letter of 6 July 1829, from Ansbach to the son Friedrich in Erlangen, with the sentence "Stay
well—you and Eduard and Ludwig—and let me hear from you
soon." Karl lived in retirement in Erlangen for 6 years and then
died quietly on 12 March 1834.

Biographers of the Feuerbach family have been unexpectedly
generous in granting space to Karl and his accomplishments in
mathematics (**12**). But there is little about him in books on mathematics beyond the aforementioned article by Cantor (**8**) and prac-
tically nothing about him in any English language source. I can
think of no more appropriate ending for this article than a very
liberal interpretation of part of a letter that was written by Karl
while he was imprisoned in the New Tower in Munich:

> I will step lightly on my tiptoes, softly, slowly, and qui-
> etly, as is customary among ghosts, and then I will wrig-
> gle through the secret little knothole I have found in the

oaken door of my cell. I will go right up to the attic and find my way there through the things that are covered up and stored away, and then slip out to freedom through a crack under the eaves. I will take a long deep breath, and then soar up and up, hither and yon, and I won't stop until I reach the benevolent moon. Then I will sit down on the very tip of the left horn of the moon, and sing a song of farewell to the earth and my beloved brothers and sisters. I will tarry a while and wait patiently for the great comet that is supposed to appear in 1825, and I hope to beg for a tiny spot in the dust of the comet's trail that shall be mine and mine alone. And then with full sails, I will fly away over all the worlds.

References

1. Material for this article was generously placed at my disposal by the following: the Hauptstaatsarchiv and the Bayerische Staatsbibliothek at Munich; the Gymnasium Fridericianum in Erlangen; the library at the University of Heidelberg; the University of Freiburg (Br.); the Zentralbibliothek at Zurich; the Bibliothéque Nationale in Paris; the New York Academy of Medicine; and the New York Public Library. I also wish to express my gratitude to J. J. Burckhardt of the University of Zurich for his help in locating references to the work of Euler in original copies of *Novi Commentarii Academiae Scientiarum Imperialis Petropolitanae* (*Petropoli*) and *Nova Acta Academiae Scientiarum Imperialis Petropolitanae* (*Petropoli*).
2. C. J. Brianchon and J. V. Poncelet, "Géométrie des Courbes: Recherches sur la détermination d'une Hyperbole équilatére au moyen de quatre conditions données," Gergonne's *Annales de Mathematiques* 11, 205 (1 Jan. 1821).
3. L. Euler, "Solutio Facilis Problematum Quorumdam Geometricorum Difficillimorum," *Novi Commentarii Academiae Scientarum imperialis Petropolitanae* (*Petropoli*) 11, 103 (1765).
4. J. Steiner, *Die geometrischen constructionen ausgefuhrt mittelst der geraden Linie und eines festen Kreises* (Dummler, Berlin, 1833), p. 40.
5. Terquem, "Considérations sur le Triangle Rectiligne," *Nouvelles Annales de Mathématiques* 1, 196 (1842).
6. Mackay, "History of the Nine Point Circle," *Proc. Edinburgh Mathematical Soc.* 11, 19 (1892).
7. J. Lange, "Geschichte des Feuerbachschen Kreises," *Wissenschaftliche Beilage zum Jahresbericht der Friedrichs Wederschen Ober-Realschule zu Berlin* Programme No. 114 (1894).
8. M. Cantor, "Karl Wilhelm Feuerbach," *Sitzungsberichte Heidelberg Akademie der Wissenschaften-Math. Naturwissen. Klasse* Abh 25 (1910).
9. L. Feuerbach, *Anselm Ritter von Feuerbach's Biographischer Nachlass* (Weber, Leipzig, 1853).

10. H. W. J. Thiersch, *Friedrich Thiersch's Leben* (Winter'she, Leipzig and Heidelberg, 1886) vol. 1, p. 251.

11. K. W. Feuerbach, Oken's *Isis* 6, 565 (1826).

12. G. Radbruch, *Gestalten und Gedanken.* Die Feuerbachs Eine Geistige Dynastie (Koehler and Amelang, Leipzig, 1948); *Paul Johann Anselm Feuerbach* (Springer, Vienna, 1934); H. Eulenberg, *Die Familie Feuerbach* (Engelhorns, Stuttgart, 1924); T. Spoerri, *Genie und Krankheit* (Karger, Basel and New York, 1952).

Index